Airborne Anti-Submarine Warfare

Airborne Anti-Submarine Warfare

From the First World War to the Present Day

Michael E. Glynn

FRONTLINE BOOKS

First published in Great Britain in 2022 by
Frontline Books
An imprint of
Pen & Sword Books Ltd
Yorkshire – Philadelphia

ISBN 978 1 39909 273 9

Typeset by Mac Style
Printed and bound in the UK by CPI Group (UK) Ltd,
Croydon, CR0 4YY.

Pen & Sword Books Limited incorporates the imprints of Atlas,
Archaeology, Aviation, Discovery, Family History, Fiction, History,
Maritime, Military, Military Classics, Politics, Select, Transport,
True Crime, Air World, Frontline Publishing, Leo Cooper, Remember
When, Seaforth Publishing, The Praetorian Press, Wharncliffe
Local History, Wharncliffe Transport, Wharncliffe True Crime
and White Owl.

For a complete list of Pen & Sword titles please contact

PEN & SWORD BOOKS LIMITED
47 Church Street, Barnsley, South Yorkshire, S70 2AS, England
E-mail: enquiries@pen-and-sword.co.uk
Website: www.pen-and-sword.co.uk

Or

PEN AND SWORD BOOKS
1950 Lawrence Rd, Havertown, PA 19083, USA
E-mail: Uspen-and-sword@casematepublishers.com
Website: www.penandswordbooks.com

Contents

List of Illustrations   vi
Introduction   viii
List of Abbreviations   xi

Chapter 1   The Airplane's Role in
Anti-Submarine Warfare   1

Chapter 2   Characteristics of the Airplane   15

Chapter 3   Characteristics of the Submarine   27

Chapter 4   Airborne ASW History: Genesis to World War II 44

Chapter 5   Airborne ASW History: The Cold War to the Present 60

Chapter 6   Airborne ASW Weapons   76

Chapter 7   Airborne ASW Sensors   92

Chapter 8   Tactical Oceanography and Environmental Analysis 114

Chapter 9   Sound Propagation Analysis   146

Chapter 10   Intelligence Gathering and Cueing   156

Chapter 11   Crew Resource Management   173

Chapter 12   Search   187

Chapter 13   Localization   205

Chapter 14   Tracking   223

Chapter 15   Attack   238

Index   251

List of Illustrations

Figure 1   Notional LOFARGram
Figure 2   Transmission Loss Curve
Figure 3   Sound Propagation under Isovelocity and Isothermal
            Conditions
Figure 4   Ray Trace of Sound Propagation
Figure 5   Shallow Water Isothermal Water Column
Figure 6   Deep Water Temperature Versus Depth Plot and Sound
            Speed Profile
Figure 7   Mixed Layer Temperature Versus Depth Plot and Sound
            Speed Profile
Figure 8   Surface Duct, Limiting Ray, and Shadow Zone
Figure 9   Convergence Zone Propagation, Critical Depth, and
            Conjugate Depth
Figure 10   Bathymetric Effect on Propagation Path Availability
Figure 11   Ambient Noise in the Ocean
Figure 12   Wenz Curves
Figure 13   Sound Speed Profile and Ray Trace
Figure 14   Transmission Loss, Figure of Merit, and Ambient Noise
            Impact on Sensor Performance
Figure 15   Transmission Loss Plot with Direct Path and Bottom
            Bounce Available
Figure 16   Notional Full Field Plot
Figure 17   Bathymetric Impact on Full Field Plot and Convergence
            Zone Availability
Figure 18   Cumulative Detection Probability Versus Time
Figure 19   Notional Local Operating Areas, Transit Lanes, and Patrol
            Areas
Figure 20   Notional Farthest-on Circle, Hazard Regions, and Choke
            Points
Figure 21   Passive Sensor and Active Sensor Areas of Probability
Figure 22   Circular Localization Pattern
Figure 23   Localization Patterns for a Line of Bearing Detection

Figure 24   Gain, Counterclockwise Bearing Drift, and Lost Contact
Figure 25   Area of Probability based on Intermittent and Continuous
            Sound Sources
Figure 26   Tracking Maneuver Techniques
Figure 27   Post A, Post B, and Additional Sensors During Event
            Tracking Scenario
Figure 28   Post A, Post B, and Post C During Event Tracking
            Scenario

Introduction

As a young anti-submarine warfare (ASW) pilot, I tried as best I could to immerse myself in the world of submarine hunting. The topic seemed at first to me to be equal parts hard science, an art developed from plenty of practice, and a set of aphorisms and rules passed down from aviator to aviator about what worked and what didn't while an aircrew was "on station."

As I learned more and practiced in the air, I found myself growing increasingly frustrated. I found myself asking questions about why a certain tactic was employed or why certain systems on the airplane developed. The tactics manuals were simply instructions, more cookbooks than explanations. Doctrinal publications were no better. They made it clear how the U.S. Navy as a whole faced off against a submarine threat, but they did not explain how the service had done so in the past or why the airborne ASW community had evolved the way it had.

New assignments came, and with them came new perspectives. Why did the British Royal Air Force (RAF) ASW aviators fly tactics differently that the Americans? "Why had their aircraft and sensors developed differently than ours?" I wondered.

During an assignment helping to develop tactical decision aid software, I was forced to immerse myself in academic works on operations analysis and oceanography. This job proved itself a tipping point. Having flown in the military for nearly a decade, I felt as though I was just starting to systematically understand airborne ASW and the fundamental relationships between the strengths and weaknesses of the airplane and submarine.

This realization was maddening. The topics might have been a bit arcane, but they were all easily grasped with a bit of attention. An aviator did not need a PhD in mathematics to understand operations analysis and did not need a master's degree in oceanography to understand the science behind submarine hunting. Reading about the holistic history of ASW gave insight into why the weapons and sensors had evolved the way they had. This was information I wished I had had a decade before. It would have made the process of learning to be a productive member of an ASW aircrew much easier and much more satisfying.

In my final flying assignment, I was sent to teach at an advanced jet training squadron. Since the instructor pilot's ready room was overwhelmingly filled with strike fighter pilots, I picked up a copy of Robert L. Shaw's *Fighter Combat: Tactics and Maneuvering*, a foundational work that breaks down the fundamentals of air-combat maneuvering. *Fighter Combat* wasn't filled with sensitive tactical secrets. On the contrary, it surveyed the basics of how airplanes maneuver and gave insight into the enduring fundamentals of aerial combat.

Shaw's work was an inspiration. There was no other work I could find that broke down the simple, enduring fundamentals of airborne ASW flying as his did for air-combat maneuvering. So, with that in mind, I set to the task of writing that book myself.

Airborne ASW is focused on countering the most powerful of weapons platforms, the submarine. Many submarine operations are very sensitive and demand security around them. In that vein, any reader hoping to gain secrets or tactical information in this book will be sorely disappointed. This work is drawn entirely from open-source academic and historical writing. It is written in such a manner to give readers insights into fundamental and enduring operational relationships for airborne ASW. We will not delve down into tactical minutiae.

Airborne Anti-Submarine Warfare is intended to offer a survey of a broad topic. As a result, we will focus our conversations on the general rather than the specific. The reader will find the first ten chapters cover the fundamentals of airborne ASW flying. The book begins with a section on the airplane's role in ASW and moves to review the strengths and weaknesses of the airplane. A brief overview of the history of airborne ASW comes next, followed by sections on weapons and sensor theory. We will move then to review tactical oceanography and environmental analysis before discussing methods of intelligence gathering and information management. Finally, we will focus on the fundamentals of ASW flying that will help pilots and sensor operators to be more effective members of their aircrew. We will spend a chapter discussing crew resource management before reviewing the five phases of airborne ASW: mission planning, search, localization, tracking, and attack.

Mission planning is a task that changes a great deal from organization to organization. As a result, it is hard to specify the exact steps that aviators ought to take while they conduct this task. I have chosen not to include a chapter for it. Rather, most of the discussion of pre-flight analysis and planning can be found in the chapter on localization. Therefore, we will discuss the airborne ASW kill chain in four steps: search, localization, tracking, and attack.

Readers should use this work in a way such that they get the most out of it. A new aviator just learning the craft will be well served to read the whole book from start to finish. However, more attention should be paid to the fundamentals sections in the first ten chapters. This will give young pilots and sensor operators a sound foundation as they learn.

Experienced aviators know what works and what doesn't from years already spent in the cockpit. They will be better served by reviewing the sections on airborne ASW history, sensor theory, weapons theory, and operations analysis in the search chapter. A review of the fundamentals will be especially useful if the aviator has been out of the airplane for some time on staff duty.

An engineer or acoustician working on an airborne ASW project will already have an excellent technical baseline to begin with. In their case, they should focus on the final five chapters. This will give them insight into what issues and problems ASW aviators deal with on station and how they might be able to help.

In writing this book, I have tried to focus as widely as possible and represent the experiences and histories of the many countries that have flown ASW aircraft. However, as an American naval officer, I invariably gravitate towards some of the language and terminology used by the U.S. Navy. I have attempted to utilize both NATO and Russian naming conventions for submarine classes, weapons, and equipment.

The largest failure of this work is a focus exclusively on fixed-wing ASW aircraft. The majority of the historical case studies in ASW flying come from an era before helicopters were used as submarine-hunting platforms. Additionally, as a maritime patrol pilot, I am ill equipped to write about Crew Resource Management (CRM) and other topics from the perspective of rotary-wing aviators. Hopefully in the future, colleagues from the helicopter ASW communities will offer their perspective and knowledge about the fundamentals of submarine hunting with rotary-wing aircraft.

As a young aviator, I wished for the ability to be able to sit down with a single book and begin to understand the basics of airborne ASW. I wanted to grasp how my airplane and my mission fit holistically with the larger ASW force. I wanted to understand the fundamentals of the craft rather than rely on tribal knowledge. It is my sincere hope that this book might do the same for my fellow aviators and ASW enthusiasts.

List of Abbreviations

AAA	Anti-aircraft artillery
ACINT	Acoustic Intelligence
AGI	Auxiliary Intelligence Gathering Ship
AIP	Air Independent Propulsion
AN	Ambient Noise
AOP	Area of Probability
ASCM	Anti-ship Cruise Missile
ASUW	Anti-surface Warfare
ASW	Anti-submarine Warfare
ASWORG	Anti-submarine Warfare Operations Research Group
C2	Command and Control
CAS	Computer Assisted Search
CDP	Cumulative Detection Probability
COMINT	Communications Intelligence
CRM	Crew Resource Management
CVBG	Carrier Battle Group
dB	Decibels
DI	Directivity Index
DSC	Deep Sound Channel
DSL	Deep Scattering Layer
EER	Explosive Echo Ranging
ELINT	Electronic Intelligence
EO	Electro-optical
ESM	Electronic Support Measures
FDS	Fixed Deployable System
FM	Frequency-modulated
FOC	Farthest-on Circle
FOM	Figure of Merit
FSASW	Full Spectrum ASW
GIUK	Greenland, Iceland, and United Kingdom (Gap)
GPS	Global Positioning System
HF/DF	High Frequency Direction Finding

HVU	High Value Unit
HWT	Heavyweight Torpedo
ICS	Intercommunications System
IDP	Instantaneous Detection Probability
IMINT	Imagery Intelligence
IR	Infrared
ISAR	Inverse Synthetic Aperture Radar
ISR	Intelligence, Surveillance, and Reconnaissance
IUSS	Integrated Undersea Surveillance System
I&W	Indications and Warnings
JMSDF	Japan Maritime Self-Defense Force
kyds	Kilo-yards
LIDAR	Light Detection and Ranging
LLOA	Limiting Line of Approach
LOFAR	Low-Frequency Analyzer-Recorder
LWT	Lightweight Torpedo
MAC	Multistatic Active Coherent
MAD	Magnetic Anomaly Detection
MANPADS	Man Portable Air Defense Systems
NATO	North Atlantic Treaty Organization
NL	Noise Level
OODA	Observe, Orient, Decide, Act
RAF	Royal Air Force
RAP	Reliable Acoustic Path
RD	Recognition Differential
RL	Reverberation Level
RWR	Radar Warning Receiver
SAM	Surface to Air Missile
SAR	Synthetic Aperture Radar
SE	Signal Excess
SIGINT	Signals Intelligence
SL	Source Level
SLBM	Submarine Launched Ballistic Missile
SN	Self Noise
SNR	Signal to Noise Ratio
SONAR	Sound Navigation and Ranging
SOSUS	Sound Surveillance System
SPA	Submarine Probability Area
SPL	Sound Pressure Level
SSBN	Nuclear Ballistic Missile Submarine

SSG	Diesel Electric Guided Missile Submarine
SSGN	Nuclear Guided Missile Submarine
SSK	Diesel Electric Attack Submarine
SSM	Midget Submarine
SSN	Nuclear Fast Attack Submarine
SSP	Sound Speed Profile
SURTASS	Surveillance Towed Array Sensor System
TASW	Theater Anti-submarine Warfare
TL	Transmission Loss
TMA	Target Motion Analysis
TS	Target Strength
TSS	Tactical Surveillance Sonobuoy
UAV	Unmanned Aerial Vehicle
USV	Unmanned Surface Vehicle
UUV	Unmanned Underwater Vehicle
XBT	Expendable Bathythermograph

Chapter 1

The Airplane's Role in Anti-Submarine Warfare

B efore we explore the history of airborne ASW and delve into how sensors and weapons are used to counter submarines, it is best to conceptualize the goals and aims of an ASW campaign. In doing so, we can understand the different roles that ASW aircraft play in such a conflict.

Modern submarines are versatile and powerful weapons. Their ability to hide themselves from observation allows them to enter defended areas and carry out missions unmolested. They often carry specialized intelligence-gathering equipment. The amount of combat power they bring to bear is highly disproportionate to their small size and relatively low cost. The submarine's combat power is great, and therefore must be countered.

At its core, ASW is negative in its aims. The ultimate goal is to enable friendly units to accomplish their missions without interference from submarines. To accomplish this goal, ASW units focus on degrading the submarine's ability to counter friendly forces. If enemy submarines are sunk during an ASW campaign, all the better. But it is crucial to keep in mind that the effectiveness of an ASW campaign is measured by how well friendly forces can accomplish their mission, not by how many hostile submarines are damaged or sunk.[1]

Rather than facing off against submarines in a symmetric, force-on-force fight, some nations have adopted strategies that attempt to attack the adversary's submarine kill chain as a whole, rather than focusing on defensive operations and targeting individual submarines. At the time of writing, the U.S. Navy refers to this type of doctrine as Full Spectrum ASW (FSASW).[2] It brings together many disparate military capabilities to attack and degrade the capabilities of submarines at every point in their journey from the pier to the point of firing a weapon.[3]

FSASW seeks to degrade the strengths of submarines and exacerbate their weaknesses. The doctrine calls for attacks against submarines in port, efforts to degrade command and control (C2) networks, strikes against land-based cueing sensors, and the use of denial and deception measures. The American FSASW concept is broken down into ten different threads, which are listed below:

Thread 1:   Create Conditions Where an Adversary Chooses not to Employ Submarines
Thread 2:   Defeat Submarines in Port
Thread 3:   Defeat Shore-Based C2
Thread 4:   Defeat Submarines near Port in Denied Areas
Thread 5:   Defeat Submarines in Choke Points
Thread 6:   Defeat Submarines in the Open Ocean
Thread 7:   Draw Enemy Submarines into Kill Boxes
Thread 8:   Mask Friendly Forces from Detection and Classification
Thread 9:   Defeat Submarines in Close Battle
Thread 10: Defeat the Incoming Weapon

In the remainder of the chapter, let us examine each thread in detail and determine what role ASW aircraft play in each phase.[4]

Thread 1: Create Conditions Where an Adversary Chooses not to Employ Submarines

Submarines bring many benefits to the battlefield. They are by their nature small, stealthy, and difficult to detect. A diesel-electric submarine, well maintained and operated by a proficient crew, is extremely difficult to detect in shallow, noisy environments close to the coast. Armed with wake-homing torpedoes and anti-ship cruise missiles (ASCMs), they possess the sensors and weapons necessary to find and sink merchant ships or warships.

Modern submarines are relatively cheap. At the time of writing, any nation with US $500 million can purchase an extremely capable diesel-electric hunter killer submarine (SSK). This is an enormous amount of capability for a small capital investment.

Submarines are by their nature asymmetric weapons platforms. Traditionally, they have been the tools of weaker navies. They have frequently been built and operated by forces that cannot challenge their better-armed opponents in a decisive fleet action and thereby seek another solution. It was the power of the Royal Navy that led Germany to develop the first U-boats in the years before World War I. So too was it the strength of the U.S. Navy's aircraft carriers and battle groups that drove the Soviet Navy to build its massive fleet of submarines during the Cold War.

To reduce the submarine threat before a conflict begins, leaders should work to convince potential adversaries that employing submarines will be futile during wartime and that doing so will strategically undermine the adversary's goals. The adversary must be convinced that they have more to lose by deploying a submarine force than they have to gain. Themes should include that submarines are expensive, difficult to operate, challenging to maintain proficiency with, and easily countered by friendly ASW capabilities.

ASW aircraft can best contribute to this thread during peacetime by maintaining a high state of readiness and demonstrating the ability to search for and track friendly and foreign submarines. Commanders should highlight the proficiency of their aircrews during exercises or real-world submarine prosecutions.[5] A strong ASW force serves as a powerful deterrent to would-be adversaries.

Thread 2: Defeat Submarines in Port

A submarine in port is usually far easier to locate than one that is underway. As such, the second thread of FSASW calls on commanders to attack boats while they are pierside, where their location is well known, they are readily observed, and may be attacked using general-purpose weapons rather than less numerous and more expensive specialized ASW weapons. Another facet of this effort is damaging or destroying submarine support infrastructure. This includes attacking headquarters buildings, maintenance facilities, weapons storage bunkers, and other support equipment. Because submarines are required to periodically return to port to resupply and rearm, degrading support infrastructure is useful to slow the resupply process. A less effective resupply and support capability gives friendly forces more time to locate and target boats in port. It also minimizes the amount of time a particular submarine is underway where it can attack friendly targets.

ASW aircraft may play a role in this thread. In the lead-up to conflict, they can help locate adversary submarines in port, using electro-optical (EO) sensors and radar. Once conflict has begun, they can utilize synthetic aperture radar (SAR) to locate boats pierside and provide targeting information to strike aircraft. Some patrol aircraft have been outfitted with land-attack cruise missiles, allowing them to strike submarines and port infrastructure. ASW aircraft may also be able to use EO and radar imagery to conduct bomb-damage assessment after a strike.

All of these missions are contingent on the strength and capabilities of adversary air defenses. Naval bases are valuable and are likely to be well defended. If hostile fighters or ground-based air defenses are active, modern fixed-wing ASW aircraft will have to remain far away due to their relatively slow speed, lack of maneuverability, and limited self-defense capabilities. The environment must be permissive for ASW aircraft to bring their sensors to bear and support attacks against port facilities.

Thread 3: Defeat Shore-Based C2

Submarines may be blessed with stealth, but this advantage comes at the price of a very limited ability to sense the tactical environment. Submarines are not very mobile when compared to surface ships or aircraft. Even nuclear submarines, with their unlimited endurance, mostly patrol at slow speeds to avoid generating large amounts of noise and becoming readily detectable. Operating slowly limits the search rates that a submarine can generate. This low search rate limits the number of targets a boat can detect and the size of the operating area it can sanitize.

Sensors such as radar or active sonar cover large areas and plot target locations rapidly and precisely. However, to remain hidden, submarines must eschew these active sensors and limit their emissions by primarily using passive sonar, visual observations, and passive electronic support measures (ESM) to orient themselves to the tactical environment. Compared to other sensors, passive sonar is ambiguous. Extensive interpretation and target motion analysis (TMA) is required to generate useful tracking data from passive sonar systems. Passive sonar may also have relatively short detection ranges, especially if the environment is very noisy or the target the submarine is searching for is quiet.

Submarines are also able to search relatively covertly by using visual means. However, because their periscopes reach only a short height above the waves, their visual detection range is limited due to line of sight. Boats can utilize ESM equipment to detect electronic signals of interest. However, this ESM equipment usually provides only directional information, not precise target location information. Since many radars and electronic signal emitters have similar signal parameters, ESM is also inherently ambiguous.

The limits of passive sonar, visual sensors, and ESM combined with the low speeds of most submarines mean that boats are limited in how rapidly and precisely they can observe the tactical environment. Captain William Toti, the author of the U.S. Navy's FSASW doctrine, captures this reality very well in the following passage:

> People generally think of submarines as having claustrophobic physical environments, and they do, but what they really have is claustrophobic

sensory environments. They are very limited in what they can perceive. Using acoustics, they generally have a maximum detection range of less than 50 miles. They must be at periscope depth to use optical systems, but detection ranges are limited by a very low periscope height of eye, and so they are generally less than 15 miles. For real long-distance targeting they need intelligence or outside cueing, both of which can be denied to them by other means.[6]

The sensory capabilities and mobility of submarines make them highly reliant on shore-based C2 and sensor networks to deliver cueing information and guide them to their targets. As a result, ASW forces can greatly degrade the effectiveness of submarines by targeting C2 networks and their associated cueing sensors. Kinetic strikes, electronic warfare, cyber-attacks, and special operations raids are all methods that a determined attacker can use to degrade C2 systems.

ASW aircraft have a limited role to play in these efforts. If properly equipped, they may be able to detect the emissions of sensors that feed the C2 network. This location information can be used by intelligence services before a conflict to plan attacks or can be fed to strike aircraft for dynamic targeting during wartime. If equipped with land-attack weapons, patrol aircraft can launch kinetic strikes of their own. However, if an adversary has even limited air defenses, it is unlikely modern ASW aircraft will be involved in strike operations until air superiority has been established.

Thread 4: Defeat Submarines near Port in Denied Areas

This step takes the ASW fight close to the enemy shore. The open ocean is often a challenging place to locate a submarine due to the sheer size of the operating areas that ASW forces must sanitize as they hunt for a hostile boat. However, an ASW force can be reasonably sure that the waters near adversary submarine bases will offer numerous chances to detect and attack boats as they go to sea or return to port. The challenge, however, is overcoming hostile defenses in these areas.

Close to shore, an adversary is likely to have a significant concentration of air defenses, surface ships on patrol, and underwater sonar arrays. Each of these capabilities makes it challenging for ASW forces to lie in wait close to port and attack submarines. As a result, forces tasked for this thread are usually stealthy or expendable.

ASW forces have typically used mines or submarines to attack boats leaving or returning to port. Nuclear fast attack submarines (SSNs) can execute offensive patrols near ports. Mines are meant to either damage hostile submarines directly or force them to transit through mine-free areas that are well known to the ASW force. This funneling effect slows down deploying submarines and leaves them vulnerable to attacks by other forces.

While some ASW aircraft are capable of deploying air-dropped mines, the high threat environment near shore precludes modern patrol aircraft from mining close to most adversary naval bases. Recently, air-dropped mines that can be deployed from long range have been developed, but they are currently only carried by bombers rather than by ASW aircraft.[7]

Further out to sea, away from hostile air defenses, ASW aircraft can be tasked to hunt down boats as they come or go from port. Often, geography and bathymetry govern the routes that submarines must take as they travel from port to their patrol areas. If transit routes are well known, ASW aircraft can be tasked to patrol these areas and attack whatever boats they come across.

While discussing this thread, we should examine a concept known as "virtual attrition." Recall that the goal of FSASW is to degrade the effectiveness of the submarine threat as a whole, not merely to sink submarines. Whatever actions ASW forces take to prevent a submarine from accomplishing its mission are ultimately beneficial. If a submarine is delayed or engaged in operations other than its primary mission, it cannot bring its weapons or sensors to bear.

Imagine an adversary that is in possession of a fleet of six submarines. Each boat has an endurance of 60 days, after which it must return to port. Resupplying, rearming, and maintenance take 30 days to complete. As such, four boats are typically at sea while two are in port. Assuming each boat takes five days to transit to its wartime operating area, we can assume each boat has 50 days of operational capacity during its patrol.

What if an ASW force could delay the transit of the submarines? By making them sail more slowly or take a circuitous route, imagine each boat must now spend 10 days rather than five days in transit. In the 90-day cycle of a patrol and refit period, the ASW force has now degraded the adversary's patrol capacity from 50 days in the operating area per submarine to 40 days. This represents a 20 percent decrease in the presence of hostile submarines in an area we wish to control. Every less day a boat is on patrol in its operating box equates to fewer opportunities for it to detect targets, launch attacks, or accomplish some other mission. If an ASW force presents a credible threat to force a change in the submarine's transit patterns, they might even accomplish this task without launching actual attacks. The key is that adversary submarine effectiveness is reduced. This degradation of submarine operational capability is the heart of virtual attrition.

Traditionally, ASW aircraft have been used extensively to patrol and attack submarines in transit lanes to and from port. Sometimes these patrols would sink boats outright, scoring a so-called seaworthiness kill. Other times, aircraft would damage a submarine so badly that the boat would be obliged to return to port. This is known as a mission kill.

Harassment from patrolling ASW aircraft often forces submarines to move more slowly and cautiously or to take a more roundabout transit route to the patrol area. Scoring mission kills or forcing submarines to take more time to reach a patrol area are both examples of imposing virtual attrition on an adversary. Operations against boats in transit to and from port are core missions of ASW aircraft. Patrol aircraft have historically played a very important role in defeating submarines in transit lanes and in imposing virtual attrition in general.

Thread 5: Defeat Submarines in Choke Points

Geography and bathymetry ensure that submarines must follow certain paths as they move to and from their operating areas. These regions where land or the ocean bottom funnels submarines are known as choke points. ASW forces can position their cueing sensors, mines, ships, submarines, and aircraft in these choke points to achieve higher rates of detection and attack than they would elsewhere.

ASW aircraft can be used to patrol choke points themselves. However, this is usually not ideal because of the relatively limited endurance of an aircraft compared to a ship or submarine. More often, patrol aircraft serve as quick-reaction assets. Perhaps a sonar array has been placed on the ocean bottom in a choke point. When the array detects a submarine passing by, operators can inform an ASW crew standing by on alert. The high speed of most patrol aircraft means they can arrive rapidly at the choke point and begin searching nearby for the submarine.

During peacetime, ASW aircraft have often been used to react to cueing in choke points and establish a track of the target submarine for an extended period of time. This extended track can be made up of many sequential flights or can consist of an ASW aircraft turning over the trail to a more persistent asset such as a frigate or SSN. During the Cold War, the U.S. Navy made extensive use of sonar arrays to cue land-based ASW aircraft to the presence of Soviet submarines.[8] Patrol aircraft would launch and begin searching in response to the cueing information. Once they found their target, aircrews would trail for some time until a friendly SSN arrived and relieved the ASW aircraft of tracking duties.[9]

These reaction and trailing missions were extremely beneficial. They provided high fidelity recordings that helped identify the characteristics, capabilities, and acoustic signature of various classes of submarines. They also provided track data that allowed planners to plot submarine transit lanes and patrol areas. They provided statistical descriptions of how submarines moved that could be used to predict the location of Soviet boats. All this information was crucial to accomplishing the sixth thread of FSASW.[10]

Thread 6: Defeat Submarines in the Open Ocean

Finding submarines in the open ocean is a challenging task. In this thread, we are faced with overcoming the greatest strength of the submarine, namely its stealth. Open-ocean search requires sanitizing large swathes of the ocean. Paradoxically, while open-ocean search and detection is a very challenging problem and a difficult part of FSASW, it is the area where many navies have invested the majority of their efforts and research.[11]

One reason that ASW practitioners have sought to attack the stealth of the submarine head-on is the action-oriented and offensive mindset of most naval officers. Sailors inhabit the tactical and operational levels of war. It is natural, therefore, that their responses to the submarine threat often focus on attacking the submarine directly rather than considering the problem of adversary submarines from a more strategic level and attacking the adversary force in a more holistic manner. ASW operations in the open ocean have often served as the stage on which this force-on-force contest plays out.

Open-ocean ASW has a distinctly economic component. Patrolling and sanitizing large areas of the ocean relatively quickly is expensive. As submarines become quieter and less detectable, the investment required to realize the tactical performance necessary to detect them grows considerably. Investing in a stealthy submarine offers an asymmetric advantage because it requires an adversary ASW force to invest disproportionately large amounts of time, effort, and resources in meeting the new, quieter threat.

Luckily, there are some factors that favor the ASW force in open-ocean search operations. First, as operations move further from land, the threat of air defenses is much lower. This makes open-ocean ASW generally safe for lightly-armed ASW aircraft. Second, the open ocean is often a much less challenging environment in which to search acoustically for submarines. Close to shore, there is often great variation in the ocean environment. In the open ocean, however, the water column is generally much more uniform. This makes search operations with sonar far simpler.

The open ocean is also generally quieter than areas closer to shore. Shipping traffic is lighter and the deep water is usually home to far less marine life. Lower levels of ambient noise make quiet submarines more readily detectable. Deep water also raises the possibility that certain long-range acoustic arrival paths such as convergence zones and the Deep Sound Channel (DSC) are possible. These arrival paths enable long-range detections that are simply not available in shallow water.

ASW aircraft perform several important roles during open-ocean ASW. These missions include searching in response to cueing information, sanitizing operating areas prior to the arrival of a high value unit (HVU), patrolling the edges of a convoy, conducting wide-area search operations, and conducting water sampling.

The ability to conduct these various missions effectively is due to the aircraft's combination of speed, altitude capability, and ability to employ multiple sensors. Modern fixed-wing aircraft can cover several hundred nautical miles per hour during transit. They can climb high enough to provide their sensors with a wide field of view or to establish line of sight with sonobuoys spread over dozens of miles. They can deploy dozens of expendable sensors, allowing them to simultaneously observe many different locations in an operating area.

While it might take a ship or submarine days to travel 1,000 miles and react to long-range cueing information, an ASW aircraft can cover this distance in two to three hours. This means that patrol aircraft will often be the first ASW forces to arrive on station. They can rapidly deploy a field of several dozen sonobuoys over a large operating area. This allows them to clear operating areas of hostile submarines and establish a safe haven prior to a friendly warship arriving.

Flying at altitude allows ASW aircraft to monitor sonobuoys many miles away. Altitude also provides an aircraft with a large field of view for radar search. Combining sensor coverage with high speed allows a patrol aircraft to realize high search rates. The sonobuoys deployed by ASW aircraft detect target submarines over much shorter ranges than the larger sonar arrays onboard ships and submarines. However, the fact that aircraft can deploy dozens of sensors over a large area compared to a submarine operating one very capable array means the aircraft is often a more efficient wide-area search asset.

ASW aircraft also act as a deterrent to submarines during crisis and wartime. Modern periscope-detection radars are very powerful and extremely capable. The high effectiveness of modern radars means an ASW aircraft is likely to detect even short exposures of a periscope nearby and attack the adversary boat. If a submarine commander hunting for targets detects the signature of an airborne periscope-detection radar, he must be far more cautious than he would be otherwise. The presence of an ASW aircraft overhead a convoy or HVU may be enough to dissuade a submarine commander from pressing an attack.

This respect for the potency of ASW aircraft is well founded. Traditionally, patrol aircraft have accounted for a large percentage of the submarines sunk in wartime.[12] If a submarine is not detected while approaching a target, it is very likely to be detected by an ASW aircraft once the boat has launched a weapon. Due to the limited mobility of a submarine and the high speed and maneuverability of patrol aircraft, it is difficult for a submarine to shake off pursuit and escape once it has been found by an aircraft. A submarine that is detected by a hostile ASW aircraft before, during, or after an attack has a very good chance of becoming a casualty itself. For these reasons, submariners are understandably wary when they detect a fixed-wing ASW aircraft overhead.

Thread 7: Draw Enemy Submarines into Kill Boxes

Earlier in this chapter we discussed the limited mobility of most submarines and the claustrophobic and ambiguous sensory environment that they inhabit. If an ASW force can cut off a submarine from its shore-based C2 network and lure its crew in by presenting an enticing target, the stage may be set to spring a trap. A real warship or a decoy emitting radar signals and a false acoustic signature may be sufficient to draw in a submarine.

Deception operations and setting traps for submarines are often most effective when the adversary submarine is unaware that a hostile force is nearby. In such a scenario, ASW aircraft are very useful. A patrol aircraft can deploy passive sonobuoys, withdraw, and loiter many miles away. When a hostile submarine is detected by a sensor in the kill box, the patrol aircraft can use its speed to rapidly close with, localize, and attack the submarine. These capabilities make ASW aircraft very useful in deception operations.

Thread 8: Mask Friendly Forces from Detection and Classification

This thread of FSASW focuses on overwhelming C2 networks and degrading submarine sensors through the use of avoidance maneuvers and deception measures. While ASW aircraft do not usually play a large role in this thread, it is useful for aviators to understand the strategy behind this facet of FSASW.

We previously discussed the effect of degrading the shore-based C2 network used to provide cueing information to submarines. One strategy to degrade a C2 network is to strike the sensors and headquarters themselves with kinetic or non-kinetic fires. In some campaigns, strikes on another nation's territory may not be politically desirable or feasible. In this case, an ASW force can use a combination of emissions control, electronic warfare, decoy employment, and maneuver to blind and overwhelm the adversary's sensors.

For example, an ASW force might deploy a set of decoys that radiate in the electromagnetic and acoustic spectrums in a manner very similar to real warships. How would the adversary be able to distinguish between sensor contacts that were truly targets and contacts that were merely decoys? The adversary would have to seek other methods to break out real contacts from the clutter of spurious tracks.

Warships can use a variety of methods to hide themselves from submarine sensors and to frustrate classification. Crews can utilize deceptive lighting methods or operate commercial surface search radars in lieu of specialized military radars. Acoustic jammers can be deployed to fill the ocean with noise that covers up the quieter signature of warships. Quieting systems such as Prairie-Masker can be used to reduce the range at which an adversary submarine can detect friendly warships.[13]

It is important to understand the tactical benefits of operating groups of friendly ships close together. The term convoy is often used when describing a group of merchant ships being escorted by warships. When discussing a group of warships sailing together, the term HVU is usually applied to the most valuable vessel while the escorting ships are referred to as a "screen."

Convoys and HVU screens are very useful in degrading the ability of submarines to detect and close with their targets. While it seems counterintuitive, a submarine patrolling in an operating area is less likely to detect a group of several ships sailing in close proximity than it is to detect any one of ships sailing independently.[14] By simply operating together in a convoy, friendly forces reduce the search effectiveness of a submarine.[15]

Even if a submarine successfully detects a group of friendly ships, it must close with the force, carry out an attack, and escape the engagement unharmed. This is a very tall order. One benefit of the convoy is that it makes airborne and surface escorts more effective. Because submarines are drawn to convoys once they are detected, escorting ships and aircraft have a higher chance of encountering a submarine during a screen patrol than they would during an independent open-ocean patrol. In this manner, escorting a convoy or HVU multiplies the effectiveness of ASW aircraft.

Additionally, convoys and HVU screens reduce the chance that a submarine can carry out successful attacks against multiple targets. When a boat launches a torpedo or ASCM, friendly escort forces are rapidly alerted. The convoy will speed up and begin to evade, opening rapidly from the submarine while the boat itself attempts to escape. As a result, it is generally difficult for most submarines to attack multiple targets during a single engagement. This limits the vulnerability of ships sailing together compared to those sailing separately.

Last, escorting ships and aircraft pose a distinct challenge to an attack submarine. They force the submarine to fight against determined defenders every time the boat carries out an attack.[16] A submarine approaching a single warship need only evade the defenses of that vessel alone. On the other hand, a boat approaching a convoy must elude the sensors and weapons of a team of ships and aircraft working together in a coordinated and synergistic manner. This is very challenging indeed.

Operating ships together in a group may make them slightly more vulnerable to detection and classification by land-based, space-based, or airborne sensors. But overall, the convoy and HVU screen concepts are force multipliers for an ASW force. Aviators should keep this fact in mind when seeking to deny a hostile submarine an opportunity to attack.

Thread 9: Defeat Submarines in Close Battle

This thread requires the ASW force to detect and defeat submarines as they approach within weapons range and launch attacks. Alternatively, friendly ships and aircraft can strike back against the submarine after the attack is underway. ASW aircraft play a key role in the close battle. They are excellent assets to deter an attack, detect a submarine as it approaches, or respond quickly in the aftermath of an attack.

As we mentioned earlier in the chapter, ASW aircraft patrolling over friendly forces are a strong deterrent to hostile submarines. The signature of a periscope-detection radar nearby forces a submarine commander to limit the amount of time he or she exposes a periscope during the approach. Infrequent or short periscope observations degrade the fire control solution, lowering the chance that the submarine's torpedoes will acquire their target and home successfully.

ASW aircraft have also been traditionally used as "pouncers," reacting rapidly and searching in response to cueing information passed from more capable sensors. During World War II, Allied warships carried signals intelligence (SIGINT) gear capable of intercepting and plotting the position of radio transmissions sent by U-boats as they stalked the convoy. Patrol aircraft overhead would copy this information and fly out over the horizon, often catching the U-boats on the surface.[17] During the Cold War, North Atlantic Treaty Organization (NATO) forces took advantage of towed sonar arrays onboard their surface ships to detect Soviet submarines at long range. ASW aircraft patrolling nearby could then be cued to these contacts far from friendly forces, before the Soviet boats could close within weapons range.

Thread 10: Defeat the Incoming Weapon

This final thread focuses on defeating the submarine's weapons once an attack is underway. ASW aircraft do not have a great role to play in this thread. If aviators happen to sight an inbound torpedo or ASCM, they can certainly alert the crews of nearby warships to take evasive action and activate defensive systems. Otherwise, patrol aircraft are mostly useful for detecting where the attack came from, closing rapidly, and beginning to search the area in the hopes of detecting the attacking submarine and retaliating with an attack.

Notes

1. This fundamental precept of ASW was well recognized by operational planners as early as World War II. Charles Sternhill and Alan Thorndike, American operations analysts, captured this very well when they wrote: "The aim of ASW is to ensure the use of the oceans necessary for military operations intended to bring about the defeat of the enemy. It is an auxiliary operation, necessary, but not sufficient, for overall success. In this sense, then, the actual aim of ASW is negative, to prevent enemy submarines from accomplishing their aim." Charles M. Sternhill and Alan M. Thorndike (eds), "Antisubmarine Warfare in World War II" (U.S. Navy Operations Evaluation Group, Report No. 51, 1946), p. 89.
2. For an excellent unclassified primer on FSASW, see Capt. William J. Toti, USN (ret.), "The Hunt for Full Spectrum ASW," U.S. Naval Institute Proceedings, Vol. 140/6/1,336 (June 2014).
3. While the FSASW name is young, this type of doctrine is not new. During World War II, Allied operations analysts built a very detailed and holistic picture of how German U-boats outfitted, trained, deployed, and conducted maintenance. The Allies sought to identify weaknesses in the entire submarine operating cycle and apply pressure at those points in order to degrade the overall effectiveness of the U-boat force. For details on operations analysis conducted by Allied ASW forces, see Brian McCue, "U-Boats in the Bay of Biscay: An Essay in Operations Analysis" (Washington, D.C.: National Defense University Press, 1990), pp. 99–108.
4. The author is greatly indebted to Capt. Toti for his excellent summary of the FSASW doctrine. Many phrases and concepts in the remainder of the chapter are drawn directly from his article "The Hunt for Full Spectrum ASW."

5. American aviators use the term "prosecution" to refer to a series of consecutive sorties searching for or tracking a single submarine, usually a foreign boat conducting a deployment away from its normal exercise areas.

6. Toti, "The Hunt for Full Spectrum ASW."

7. Bill Sweetman, "U.S. Discloses new Airborne Mine Concept," *Aviation Week* (24 April 2015).

8. Owen R. Cote, Jr. "The Third Battle: Innovation in the U.S. Navy's Silent Cold War Struggle with Soviet Submarines" (Newport, RI: Naval War College Press, 2003), p. 51.

9. Ibid., pp. 51–2.

10. U.S. Navy Cold War tactical decision aids used for ASW search planning used statistical databases that described the manner in which submarines moved. For example, a database on speed would show that a particular class of boat would transit at a particular speed a certain percentage of the time or maintain a certain depth for a particular percentage of the time. Extended trailing offered a large sample size and copious amounts of data to help populate these statistical databases. For details on U.S. Navy tactical decision aids, see Chapter II of Daniel H. Wagner, "Naval Tactical Decision Aids" (Monterey, CA: Naval Postgraduate School, 1989).

11. Toti puts it well when he describes focusing primarily on overcoming the submarine threat by concentrating on defeating its strongest asset, open ocean stealth, as "an anti-Sun Tzu approach." Toti, "The Hunt for Full Spectrum ASW."

12. During World War II, ASW aircraft accounted for the destruction of 250 German U-boats alone and an additional 37 acting in concert with surface ships out of a total of 635 submarines lost to hostile action or unknown causes. This accounts for 45 percent of all sinkings. When compared to the fact that surface ships account for 264 of the U-boats lost, it becomes clear that ASW aircraft have historically proven very deadly to submarines. Axel Niestlé, *German U-Boat Losses During World War II* (Annapolis, MD: U.S. Naval Institute Press, 1998).

13. Prairie-Masker is a system developed by the U.S. Navy to reduce radiated noise levels of warships. Air bubbles are injected into the water near the hull to reduce the amount of acoustic energy that couples to the water. Similarly, air bubbles are injected into the water near the ship's screws to reduce the radiated noise. As a result, the overall radiated noise of the ship is lower and the vessel is less detectable to submarines nearby.

14. Jan S. Breemer, *Defeating the U-Boat: Inventing Anti-Submarine Warfare* (Newport, RI: U.S. Naval War College, 2010), pp. 62–3.

15. It is worthwhile mentioning that warships or convoys operating in close proximity together are more detectable by certain land-based, space-based, or airborne sensors. Because civilian shipping is highly unlikely to operate in the type of screen arrangements that warships often employ, these sensors can identify the group of ships sailing together. The best FSASW strategies will work first to degrade cueing sensor networks while also using defensive measures like convoys or HVU screens to degrade the search efforts of individual submarines.

16. The necessity for a submarine to fight for every kill it claims against a convoy is very well summarized by the historian Jan Breemer in his work on the German submarine campaign and Allied response during World War I. Breemer, *Defeating the U-Boat*, p. 63.

17. Bryan Clark and John Stillion, "What it Takes to Win: Succeeding in 21st Century Battle Network Competitions" (Washington, D.C.: Center for Strategic and Budgetary Analysis, 2015), pp. 9–10.

Chapter 2

Characteristics of the Airplane

Examining the strengths and weaknesses of fixed-wing aircraft helps us to better understand their role in ASW. Much like the submarines they hunt, the performance and capabilities of ASW aircraft have changed a great deal during the last century. However, certain inherent characteristics remain unchanged. In this chapter, we will consider the mobility, sensor coverage, and endurance capabilities of fixed-wing aircraft. We will consider how unit cost and self-defense capabilities drive the employment of these aircraft. Lastly, we will discuss how the experience of ASW aviators compares with that of the submarine crews they face off against and how this balance may change during wartime.

Mobility – The Balance of Speed and Maneuverability

Mobility is an aircraft's greatest strength, and what makes it useful for carrying weapons and sensors. High speed allows an aircraft to takeoff and fly to a search area quickly. Once it arrives, the aircraft can use its speed to rapidly cover wide swathes of the search area. When a target is detected, a fast aircraft can quickly close with the submarine to drop more sensors or employ a weapon. While not as critical a factor as speed, the maneuverability of aircraft allows them to quickly reposition themselves in three dimensions and to deliver sensors and weapons very accurately.

High speed makes fixed-wing ASW aircraft natural long-range patrol assets. Armed with cueing information, they can cover hundreds or thousands of miles in several hours. They are able to get sensors in the water rapidly before cueing information becomes "stale." In several hours, an aircraft can cover a distance that would take a submarine or ship days to transit.

High speed provides fixed-wing aircraft with a measure of protection against attack. A fast aircraft can use speed and maneuvers to evade threats from enemy fighters. The ability to rapidly open from a submarine and climb to high altitude makes fixed-wing aircraft much more challenging to target with surface-to-air weapons when compared with slow and low-flying ASW helicopters.

Speed does not come without a price. First, an aircraft's payload is much more limited than that of a ship or submarine. Second, an aircraft's endurance in a patrol area is usually only several hours, compared to days or weeks for a submarine, ship, or unmanned vehicle. Fixed-wing aircraft cannot fly below certain speeds without sufficient airflow, and the lift it generates, breaking down – especially while maneuvering. This forces them to continuously move relative to a target submarine. An aircraft having to constantly reposition is not always desirable from a tactical standpoint.

The speed and maneuverability of an ASW aircraft are strongly influenced by the design of its wing. A thick, straight wing is the best choice for low-altitude maneuverability and slow-speed flight. Both of these characteristics are very desirable to allow an aircraft to turn tightly, precisely place weapons and sensors, and to stay close to a submarine. However, a thick, straight wing is very inefficient for fast flying. To maximize transit speeds to and from a patrol area, a thin, swept wing is best. Unfortunately, this type of wing does not perform well at low speed or when turning tightly. Engineers must make compromises between these two extremes when choosing a wing design configuration for an ASW aircraft.

The ability to rapidly reposition to a new patrol area is very useful when searching for a submarine. Cueing information such as intercepted radio transmissions can reveal the presence and approximate location of a submarine. However, as soon the transmission ends, the area where the submarine may be begins to expand. The slower an aircraft is to transit the distance between its base and the search area, the greater the area it will have to clear once it arrives. The larger the area, the larger the number of sensors or the longer the amount of time needed to sanitize it. The advantage of searching with "fresh" rather than "stale" cueing data makes it highly desirable that an ASW aircraft be capable of high-speed cruising flight.

The need for a high-speed wing must be balanced with the desire for good low-speed maneuvering qualities. Generally, an aircraft optimized for high-speed flight has a high wing loading, a measure of the relationship between the size of the wing and the weight of the aircraft. A low wing loading is most desirable for tight turns and good low-speed handling. Speed and maneuverability are opposed, meaning that engineers have to make careful decisions when designing an aircraft or modifying an existing type for ASW flying.

Sensor Coverage

Compared to other types of aircraft, fixed-wing airplanes have the ability to fly at higher speeds and reach higher altitudes. These two capabilities result in fixed-wing airplanes being able to achieve sensor coverage over larger areas compared to other ASW platforms. Speed enables fixed-wing aircraft to generate high search rates. High-altitude capability allows ASW aircraft to keep line-of-sight contact with sensors spread over a wide area.

ASW aircraft can generate high search rates by one of two methods. The first is marrying their high speed with sweeping-type sensors such as radar or visual detection. A lookout in an aircraft or a radar antenna can view some particular width of the ocean as the aircraft flies through the search area. Inside a certain range, the aircrew can be reasonably confident that they will detect a submarine periscope or mast. This "sweep width" is a function of several factors, including the size of the target, the capabilities of the sensor, and the environment.

Because an aircraft can fly at high speeds, it can drag this swept area across the ocean surface quite quickly. This results in a high search rate when compared to other platforms. The detection range of a submarine or surface-ship sonar might be longer than the sweep width of the aircraft, but the speeds of these platforms are much lower than that of the aircraft. This means the aircraft will generally be able to realize higher search rates.

The other way in which aircraft realize high search rates is by using their speed to spread sensors over a large area. Ships and submarines generally search using one or several sensors located on or very near their hulls. An aircraft, however, can drop dozens of sonobuoys spaced many miles apart. High speed allows an aircraft to rapidly deploy a sonobuoy field that covers thousands of square miles. While each of the sensors is less capable than those aboard a ship or submarine, the synergistic effect of the sonobuoys searching together allows an aircraft to search large areas of the ocean very quickly. In both cases, speed is the factor that allows a fixed-wing aircraft to generate high search rates.

The other characteristic of fixed-wing aircraft that allows them to achieve wide sensor coverage is their ability to climb to medium and high altitudes. Altitude enables a sensor to observe objects at long range, because the curvature of the earth no longer blocks the sensor's field of view. It also allows aircraft to keep line-of-sight communications with sensors that are many miles away. Because an aircraft can climb to altitude, it can monitor sonobuoys spread over dozens of miles. In comparison, a surface ship's ability to keep contact with remote sensors is limited by the height of the antennae on its mast.

By combining the speed necessary to rapidly deploy sensors and the altitude capabilities that allow contact over many miles, ASW aircraft are able to seed and monitor sonobuoy fields that are spread over very large areas. While fixed-wing aircraft must make sacrifices in other mission areas, their ability to generate sensor coverage over large areas and realize high search rates gives them important advantages in ASW.

Endurance

Lower endurance is the price fixed-wing aircraft pay in exchange for their speed. Rather than days or weeks on patrol, aircraft are limited to several hours in an operating area. Endurance is a function of the amount of fuel and number of expendable sensors carried by an aircraft. It is also influenced by the design of the aircraft and the distance between the operating area and the aircraft's base.

Military aviators typically try to fly an aircraft as efficiently as possible given their tactical objectives. This will drive pilots to fly in various manners depending on what type of mission they are conducting and the current phase of the mission. The tactical objectives of an aircrew will influence how they fly during a patrol, which in turn will influence their endurance on station. An aircrew seeding a search area with sonobuoys will fly at maximum-range airspeed, trying to lay the pattern of search sensors quickly and efficiently. An aircrew conducting a visual or radar search will also fly at maximum range airspeed, to generate a high search rate while not consuming fuel too quickly. An aircrew that is monitoring a field of sensors or patrolling over a convoy or HVU will linger and fly more slowly, maintaining their maximum endurance speed. Here, minimizing fuel consumption is the concern, and flying slowly allows the aircrew to maximize their time overhead.

The amount of fuel consumed during transit and during patrol is determined by the type of powerplant and the wing design of an ASW aircraft. Historically, most ASW missions have found aircraft transiting to and from their patrol area at medium or high altitude and flying at low altitude during patrols. This creates conflicting requirements between a design optimized for high-speed cruising flight and one optimized for low-altitude endurance and maneuverability.

The ideal powerplant and wing design configuration for low-altitude endurance is a propeller-driven, piston-engine aircraft with a straight wing. These features result in excellent on-station endurance and low-speed maneuverability. However, they also result in low cruising speeds to and from the patrol area. Piston engines are also obsolete in modern applications due to high maintenance requirements, low reliability, and insufficient power densities.

The ideal configuration for high-altitude flight is a turbofan-equipped aircraft with swept wings. Turbofans are very efficient at high altitude and swept wings create low drag, resulting in fast cruise speeds and moderate fuel consumption. However, jet engines burn a great deal of fuel at low altitude, where most ASW flying has historically taken place. Also, a thin swept wing requires fast airflow over it to generate sufficient lift, especially when maneuvering. A swept-wing jet aircraft usually cannot turn nearly as tightly as a straight-wing aircraft. A lower rate of turn puts an aircrew at a disadvantage, as they cannot rapidly reverse course to launch sensors or weapons near a submarine.

For the first half-century of ASW, patrol aircraft featured piston engines and straight wings. As turbine engines came of age after World War II, ASW aircraft adopted new types of powerplants known as turboprops. The turboprop engine, where a jet turbine drives a propeller, allowed faster transit speeds than a piston engine and better fuel economy for low-level flying compared to a jet engine. In the modern era, some ASW aircraft have been built as modifications of commercial airliners, sporting jet engines and wings optimized for high-speed flight. They trade some low-altitude maneuverability and endurance for fast high-altitude transit speeds.

When discussing endurance in a patrol area, it is generally not useful to make blanket statements about what certain types of wings or engines are optimal for ASW flying. The amount of time that an aircraft can remain in a patrol area is heavily influenced by the distance between the aircraft's base and the search region. Let us consider two scenarios: first, where the patrol area is a short distance from an airbase, and the second, where the patrol area is far from the airbase.

In the first case, where the distance from takeoff to the patrol area is short, a slower turboprop aircraft will typically be able to remain on station longer than a faster turbofan aircraft. Because the distance from the base to the patrol area is short, the difference in time for the slower turboprop and the faster jet to arrive is minimal. The lower fuel burn of the more efficient turboprop means this aircraft will be able to stay on-station longer in this scenario.

In the second case, where the distance between the airbase and the search area is large, the turbofan aircraft has a distinct advantage. The slower turboprop will take a great deal of time to reach the patrol area, while the faster jet will arrive quickly. By the time the turboprop arrives, it will have consumed a large fraction of its fuel load, giving it a shorter time to search.

Conversely, the faster jet will generally be able to remain on station longer before departing. In considering these examples, it becomes clear that operational factors are just as important as aerodynamics and thermodynamics when it comes to choosing a design configuration for a patrol aircraft.[1]

For the first seven decades of ASW, fuel was the limiting factor that drove the on-station endurance for patrol aircraft. Crews relied on spotting submarines visually until the middle of World War II, when ASW aircraft began to use airborne radar. Crews might become fatigued and the radar might malfunction, but the driving factor that limited an ASW patrol was how much fuel was available.

The 1950s witnessed diesel-electric submarines fully adopt the snorkel and the introduction of nuclear-powered submarines. These two innovations allowed submarines to shift from operating primarily on the surface to operating primarily underwater. ASW aircrews shifted from searching with their eyes and radar and increasingly used expendable acoustic sonobuoys to detect their targets. With a limited number of sonobuoys onboard, crews faced the possibility of running out of sensors before they ran out of fuel. However, the high radiated noise levels of early Soviet nuclear submarines meant that running out of sonobuoys was unlikely. Fuel remained the primary factor that limited the on-station endurance of Western ASW aircrews until the mid-1980s.[2]

When Soviet submarines began to achieve high levels of quieting in the late 1970s, the acoustic advantage enjoyed by the West up to that point began to erode rapidly.[3] As the noise levels of Soviet submarines dropped, so too did the range at which these boats could be tracked with sonobuoys. A sonobuoy might track an older Soviet submarine over a dozen miles. A newer, quieter boat might only be detected over several thousand yards. Western crews were forced to use larger numbers of sonobuoys to search for and locate a target. Once the target was detected and localized, a larger number of sonobuoys had to be dropped each hour to maintain contact.[4] If the environment were noisy and the target quiet, sonobuoy consumption would be high enough that an aircraft would run out of sensors before it ran out of fuel. As submarine quieting continues to improve, the rate of sensor expenditure will likely become increasingly important to ASW aircrews in determining on-station endurance.

The endurance of an aircraft on station is the result of the many variables we discussed above. At long ranges, fuel may remain the limiting factor driving endurance. For aircrews tasked to find quiet targets in noisy, shallow waters, sensor payload and the rate of sonobuoy expenditure may determine how long an aircraft can remain on patrol.

Cost

Historically, ASW aircraft have been cost-effective options to combat submarines relative to larger platforms. When compared to a surface ship or submarine, a patrol aircraft is cheaper to build and operate. It also requires a smaller number of crewmembers, putting fewer lives at risk during combat.

Aircraft have grown increasingly complex over the last century. The cost and sophistication of the sensors used for ASW operations have increased as well. However, when compared to surface ships, patrol aircraft remain relatively inexpensive. During World War II, fifteen long-range patrol aircraft could be built for the cost of a destroyer. Today, an air arm can buy eight of even the most expensive ASW aircraft in the world for the cost of a guided-missile destroyer.[5]

Patrol aircraft also require fewer personnel to operate than do warships or submarines. Whereas a nuclear submarine may boast a crew of 100 sailors and a destroyer requires a crew of over 300, an ASW aircraft typically goes flying with fewer than 12 aviators. Aircraft squadrons still require dozens of mechanics and support personnel. But these support personnel remain ashore or on a carrier, rather than accompanying the warfighters into harm's way. A ship and a submarine bring all their support personnel with them to allow a team of several technicians to hunt. Compared to a ship, patrol aircraft put far fewer people at risk. The human cost of lives lost when an ASW aircraft is shot down is usually far lower than when a destroyer is sunk during an anti-submarine patrol.

Self-Defense and Survivability

One advantage of patrol aircraft is their relatively high levels of survivability compared to ships and submarines. This survivability is partially the result of the speed, maneuverability, and self-defense capabilities of the aircraft themselves and partially due to the relationship in space and domain between the aircraft and the submarine.

Aircraft are inherently mobile compared to other platforms. Their speed and ability to maneuver rapidly in three dimensions makes them difficult to target. This complicates the task of an attacker, regardless of whether their weapon is a submarine, ship, aircraft, anti-aircraft artillery (AAA) round, or guided missile. Speed also allows an aircraft to evade defenses and at times, withdraw to avoid an attack.

However, the true driving factor that makes aircraft less vulnerable compared to ships and submarines is that submarines' primary mode of operation often makes it difficult to attack an aircraft. A submarine operating beneath the water can easily engage another submarine or a surface ship with its primary weapon, the heavyweight torpedo (HWT). The submarine can remain in its primary operating mode, preserve its stealth, and attack a target unawares. Tracking and attacking an ASW aircraft is a much more difficult task, one that until recently has required a submarine to expose itself to operating on the surface or at periscope depth.

The earliest ASW aircraft were very primitive and vulnerable. Constructed of wood and fabric, they were underpowered and very slow. They were so slow in fact that if their crew was lucky enough to spot a surfaced submarine, the boat was usually able to easily submerge before the aviators could close to carry out an attack. Fortunately, this kept these early patrol aircraft safe, as submarine commanders would generally submerge to regain their stealth rather than stay surfaced and engage the approaching airplane.

During the interwar period, air arms began to equip their ASW forces with converted bombers and transports, many of which were heavily armed with defensive machine-gun turrets or forward-firing cannon. Aircraft such as the Consolidated B-24 Liberator and Short Sunderland packed enough of a punch to beat back attacking fighters and to suppress the AAA mounted aboard submarines. Compared to their predecessors from World War I, they were also significantly faster. This gave them a much better chance of spotting a submarine on the surface and closing to weapons range before the submarine could dive and escape.

The relative size of a submarine and an aircraft resulted in it being very unlikely that an aircraft would fly close enough to the surfaced boat to become vulnerable to attack without the aviators noticing. An aircrew could stay out of range of a submarine if they wished. They would only come within range if they were intent on attacking. When they did, they could use their forward-firing guns to strafe the submarine's deck, killing the sailors manning the deck guns and suppressing the AAA. If the aircraft did not want to engage, it could simply remain clear of the boat and out of range. As long as submarines operated primarily on the surface, the inherent mobility of the aircraft allowed the aviators to dictate the tactical scenario and remain safe from attack.

With the introduction of the snorkel and nuclear power in the 1950s, submarines shifted to operating primarily underwater rather than on the surface. This revolution made aircraft even safer due to the claustrophobic nature of the underwater sensory environment. A submarine couldn't employ AAA underwater and had no realistic way track an aircraft and to guide a missile towards the airborne target. A submarine might be able to detect the presence of a patrol aircraft overhead by its sound signature but plotting the location of the aircraft was not feasible.[6]

This permissive threat environment led to patrol aircraft discarding their defensive gun turrets and armor. In the decades after World War II, most new ASW aircraft were built as modifications of airliners or transports. Their self-defense capability was due entirely to their mobility and the submerged mode of operations of their targets. These aircraft were very vulnerable to shipboard air defense systems or marauding fighter aircraft.[7] During the 1980s and 1990s, some patrol aircraft found themselves operating in anti-shipping roles or flying in littoral environments, close to, or even over, land. These were areas where many types of air defense threats might be active. This led to the adoption of countermeasures such as chaff, flares, or towed radar decoys.[8] Some aircraft even began flying with air-to-air missiles, in case they came across an adversary patrol aircraft during a mission or had to perform a last-ditch defense against an attacking fighter.[9]

In recent years, certain nations have begun testing air defense systems that would allow a submarine to engage a helicopter or aircraft while remaining underwater. Using visual sightings from a periscope, the submarine can launch a surface-to-air missile (SAM) in the direction of the target. When the missile clears the water, its seeker head searches for and acquires the aircraft. This raises the danger to ASW aircraft.

However, fixed-wing patrol aircraft have enough mobility that they remain challenging targets. They are difficult to acquire through a periscope and their speed means they can rapidly close within and then withdraw out of the range of submarine missile systems. While these submarine air defense systems do raise the risk to patrol aircraft, they are much more threatening to helicopters. Helicopters are much slower than fixed-wing aircraft and frequently stop to hover and employ their dipping sonar. A stationary target is much easier to engage compared with a moving ASW aircraft.

Despite the emergence of new threats, fixed-wing aircraft remain generally quite survivable. They are much harder for a submarine to detect, track, and engage compared to ships or submarines. When they are destroyed, the loss of life and capital is much less compared to the loss of a larger platform. Aircraft can hunt and threaten submarines largely without exposing themselves to danger. This is not true, however, in areas where significant air defenses exist. Despite their countermeasures, modern ASW aircraft are not able to safely operate where enemy fighters or SAMs are present.

Crew Experience

The final characteristic of fixed-wing ASW aircraft we will discuss is the experience and seniority of their aircrew compared with that of the submarine. Because aircraft are generally more numerous than submarines and are flown by several crews, they are commanded by much more junior personnel. In most navies, a modern submarine is commanded by a mid-grade officer with 15 years of experience. A patrol aircraft is typically commanded by a junior officer five to ten years less senior than the commanding officer of the submarine.

This situation results in a disparity in the experience level of the officer commanding the submarine and the officer commanding the aircraft. This imbalance will often have a direct effect on the outcome of an engagement between these two platforms. Consider the case of an American ASW crew matching up against an American submarine. Frequently, the aircraft will be commanded by a junior officer under the age of 30 with less than three years of ASW experience. The submarine, on the other hand, will be command by an officer almost a decade older than the aviator and with 15 years of experience in his or her warfare area. Which platform will likely hold the advantage based on experience and knowledge?

That much being said, the experience of aircrews is highly dependent on the manning and training policies of individual services. Some services emphasize frequent personnel turnover, shuffling aviators between assignments to gain breadth of experience. Other services keep their aircrews flying with the same squadron or even with the same crew for many years. These more stable aircrews develop very deep experience in ASW and close familiarity with each other. A veteran crew who knows each other's habits intimately and are deeply familiar with ASW will probably have an advantage when their experience is compared to that of the submarine commanding officer. The balance between the tactical acumen of hunter and hunted is due in large part to personnel policies.

The balance of experience between aircraft and submarine commanders may shift during wartime. Long submarine campaigns result in the need to train large numbers of replacement crews. These crews take the place of personnel rotating ashore, exhausted after several submarine combat patrols. Or new crews may be formed to take new boats to sea as the fleet suffers casualties and builds replacement submarines.

A submarine force that is suffering relatively low casualty rates can grow in effectiveness and train increasingly more skilled and resourceful commanders. Such was the case of the American submarine force in the

Pacific during World War II. Ineffective commanders were relieved, skilled skippers molded younger officers, and the force became more lethal as the conflict wore on. On the other hand, submarine forces that suffer high attrition rates lose large numbers of their most skilled and knowledgeable commanders, often before these aces can groom the next generation of captains. Such was the case of the German U-boat force in World War II. As losses mounted, new units went to sea with increasingly lower levels of experience and enjoyed less success in combat.

The relative invulnerability of aircraft to submarines means that patrol aviators, on the whole, can continue to fight and learn over months, if not years. These crews can gain high levels of experience and learn valuable lessons that they can pass on to their compatriots. Passing on lessons learned to other aircrews is critical in wartime, due to the small number of detection and attack opportunities that present themselves.

Statistically, an ASW aircrew detecting and attacking a submarine in wartime is an improbable event. Many aircrews in World War I and World War II flew months of patrols without so much as sighting an adversary submarine, to say nothing of attacking and sinking one. Often, an aircrew might have only one chance to sight and attack a submarine during the entire war. This meant crews weren't able to fail during an attack and then apply what they had learned to improve the next time a submarine was sighted. Aircrews had to rigorously rehearse while waiting for their chance to strike, as they were unlikely to get another chance to kill an adversary boat.

Notes

1. To better understand the influence of operational factors on ASW aircraft configuration, readers should study the history of the design decisions that led to the construction of the British Nimrod MR1. An excellent discussion can be found in Chris Gibson, *Nimrod's Genesis: RAF Maritime Patrol Projects and Weapons Since 1945* (Crowborough, East Sussex: Hikoki Publications, 2015).

2. The extent to which this balance between fuel and sensors held true for Soviet ASW aircraft is unclear. Very little open source literature is available that discusses Soviet air ASW operations against NATO or Western submarines. It is reasonable to assume that because Western boats achieved high levels of quieting far earlier than Soviet designs maintaining contact with these submarines required many more sensors for a set time period compared to noisier Soviet boats. The extent to which sonobuoy expenditure rate drove mission endurance for Soviet aircrews is unclear.

    The quieter a boat, the more closely spaced the sonobuoys an aircraft will have to deploy to maintain contact. Since the number of sonobuoys carried by an aircraft is generally fixed, spacing sensors more closely together will increase the rate at which the sensors are dropped. For a discussion of the decrease in target radiated noise levels and more tightly spaced sonobuoy tactics, see Tony Blackman, *Nimrod: Rise and Fall* (London: Grub Street Publishing, 2013), pp. 74–5.

3. Owen R. Cote, Jr., *The Third Battle: Innovation in the U.S. Navy's Silent Cold War Struggle with Soviet Submarines* (Newport, RI: Naval War College Press, 2003), pp. 69–70.

4. Blackman, *Nimrod: Rise and Fall*, pp. 74–5.

5. Consider the inflation-adjusted prices of a $4.79 million Consolidated B-24 Liberator and the $69.4 million *Fletcher* class destroyer bought in 1943. Today, the flyaway cost of a Boeing P-8A aircraft is $171.6 million, compared to the $1.48 billion cost of an *Arleigh Burke* class guided-missile destroyer. George C. Wilson, "A New and Costly Warship has its Critics," *Washington Post* (24 August 1986). Department of Defense, "Budget Line Item Justification: Shipbuilding and Conversion" (February 2015). Department of Defense, "Budget Line Item Justification: Combat Aircraft, Navy" (February 2015).

6. For details of submarines detecting over-flying patrol aircraft, see Blackman, *Nimrod: Rise and Fall*, p. 75.

7. In the mid-1980s, the U.S. Navy adopted a doctrine known as the Maritime Strategy, which centered around Carrier Battle Groups (CVBGs) operating close to the shores of the Soviet Union in the Norwegian Sea and near the Soviet Far East to hold critical targets at risk during a campaign between NATO and the Warsaw Pact. Land-based patrol aircraft were critical to help screen the CVBGs, but these aircraft were very vulnerable to Soviet interceptors. In a symbiotic relationship, the CVBGs required the presence of ASW aircraft to ward off submarine attack and the patrol aircrews required defensive cover from carrier-based fighters and airborne early warning aircraft. For more information on the Maritime Strategy, see John P. Hattendorf, *The Evolution of the U.S. Navy's Maritime Strategy, 1977-1986* (Newport, RI: Naval War College Press, 2004).

8. During the Gulf War, British Nimrod MR2s were equipped with radar jamming pods, a towed radar decoy, and a self-defense suite to permit operations close to shore and over land where fighters and air defenses were likely to be active. For details on this equipment, see Blackman, *Nimrod: Rise and Fall*, pp. 21 and 127. In a similar manner, portions of U.S. Navy P-3C fleet were equipped with missile approach warning systems, chaff, and flares as part of the Anti-surface Improvement Program during the late 1990s: David Reade, *The Age of Orion: The Lockheed P-3 Story* (Atglen, PA: Schiffer Military/ Aviation History, 1998), pp. 54–6.

9. During the Falklands Campaign, British Nimrod MR2s were equipped with AIM-9 Sidewinder air-to-air missiles in case they encountered Argentinian Boeing 707 surveillance aircraft. Nimrod crews also conducted training known as "fighter affiliation" to gain a capability to force an overshoot of an attacking fighter aircraft and to employ their AIM-9s in a last-ditch defense if prior evasive action had failed. While P-3As operated covertly by the CIA over China in the early 1960s were modified to carry the AIM-9, the U.S. Navy maritime patrol force never carried these weapons operationally. Tests with P-3Cs and AIM-9Ls were carried out in the late 1980s and early 1990s, but the operational patrol crews never carried the missile. For a discussion of Nimrod fighter affiliation training, see Blackman, *Nimrod: Rise and Fall*, pp. 123–5. For information on P-3 Sidewinder integration, see Reade, *The Age of Orion*, pp. 46 and 104–05.

Chapter 3

Characteristics of the Submarine

L ike aircraft, submarines have advanced a great deal since their debut in combat over a century ago. However, some of their characteristics have remained constant. Submarines retain some inherent advantages and disadvantages when compared to ships and aircraft. Their ability to submerge hides them from sight and the reach of many types of sensors. This gives rise to stealth, their greatest strength. Operating underwater also poses challenges. Limits of endurance, communications, and self-defense cause key vulnerabilities.

The factors of stealth, mobility, and endurance are tied together for a submarine. Each factor strongly affects sensor performance and weapons employment, both for the submarine and for any aircraft that are trying to detect it. To make the best use of their aircraft and their sensors, aviators ought to understand the operational characteristics of the submarine.

Stealth

Stealth is the defining attribute of the submarine. The ability to operate underwater cloaks a boat from many methods of detection. As far as most sensors are concerned, the ocean is very opaque. Water rapidly absorbs and scatters electromagnetic waves, making many non-acoustic sensors far less effective.

Water is a very good conductor of sound energy, and it is here that modern submarines are most vulnerable. However, submarines still retain great advantages. The ocean can be very noisy, with sounds from sea life, rain, waves, and shipping traffic drowning out the signature of a quiet submarine. The water column is never constant, as variations in temperature, salinity, and currents disrupt sound propagation. To make submarines as stealthy as possible, nations have invested great sums of money into mastering design features and construction techniques that minimize a boat's radiated noise.

When submarines were first introduced to front-line naval service prior to World War I, their stealth was due entirely to their ability to operate underwater. ASW forces had no reliable method to detect a submarine once the boat descended below the surface. By the end of World War I, several nations had begun to develop hydrophones for detecting submerged targets. However, these systems were rudimentary and their tactical utility was low.

In the interwar period, important advances were made in passive and active sonar technology and in miniaturizing radar sets to allow them to be carried aboard ships and aircraft. During the first few years of World War II, however, these sonar systems were very limited in their ability to search large areas of the ocean. They had to be cued to a small area to detect and track a submarine. The primary methods of detecting submarines were visual search and radar detection. Once a submarine had submerged, it could then be tracked and attacked using active sonar. By the end of World War II, radar had advanced to the point where submarines were forced to adopt snorkels and operate primarily underwater. This shift to operating underwater drove ASW forces to exploit the acoustic vulnerability of submarines.

In the years after World War II, the introduction of the snorkel for conventional submarines and the birth of nuclear power moved the ASW struggle into the acoustic realm. As a result, submarine builders invested greatly in minimizing the radiated noise levels of their boats. Machinery and drive trains became increasingly quiet and were eventually mounted on noise-damping isolation mounts known as "rafts." Increasingly quiet boats eroded the advantages of passive sonar, driving ASW forces to adopt powerful active sonar during the later years of the Cold War. This in turn drove submarine builders to outfit their boats with features that made them less detectable to active sonar. One such method was sheathing the metal hull of the boat with a noise-absorbing rubber cover known as anechoic tiling.

Today, the stealth of a submarine is due in part to the effect that operating underwater has on non-acoustic sensors. The remainder is due to the difficulty in picking out the limited acoustic signature of the submarine from the noise and variability of the ocean itself. Acoustic stealth is closely tied to the type of powerplant used in the submarine and the manner in which the boat is operated.

Conventionally-powered submarines are usually driven by an electric motor linked to a bank of batteries. Typically, a diesel engine is used periodically to recharge the batteries as well as provide propulsion power on the surface. When operating on electric power alone, conventional submarines are generally extremely quiet. However, they must periodically expose a snorkel above the surface to take in oxygen and discharge exhaust from the diesel engine while the batteries are recharged. By raising part of the boat above the surface, the submarine is vulnerable to detection by visual search, radar, infrared (IR) sensors, and to a limited extent, chemical detectors.

The frequency at which a submarine exposes a periscope or snorkel is known as the "indiscretion rate." In the last several decades, certain nations have attempted to reduce this vulnerability by equipping their conventional submarines with propulsion systems that can run or recharge batteries without oxygen. These so-called Air Independent Propulsion (AIP) systems reduce indiscretion rate from hours and days to potentially weeks at a time. This does a great deal to reduce the vulnerability of a submarine and in doing so, increase the boat's stealth.

Unlike a conventional submarine, a nuclear-powered submarine has no need to surface to ventilate its spaces or recharge batteries. A SSN can operate submerged indefinitely, generating power and controlling its environment for as long as is necessary. As a result, nuclear-powered submarines are far less vulnerable to detection by visual search or radar. However, their ability to operate underwater comes with a price. The large amount of power generated by the reactor core also creates large amounts of waste heat. To dispose of the heat from the reactor, powerful pumps circulate water or liquid metal coolant through the core. A nuclear-powered boat must constantly run relatively noisy machinery, increasing the amount of sound energy it puts out into the water column.[1]

The size of conventional and nuclear submarines also has important effects on stealth. Conventionally-powered boats are typically smaller than their nuclear-powered brethren. Their diminutive size makes them more difficult to detect with active sonar. They are also more maneuverable, making it easier for them to operate in shallow water and hide in the clutter created by sound energy bouncing off the ocean bottom and objects in the water. The large amount of machinery required for a nuclear plant makes a SSN far larger than a SSK. A larger hull results in a more powerful return on active sonar. In general, nuclear submarines are deep-water boats, much better suited to operating off the continental shelf where they can dive deep and use their speed to evade pursuit.

Nuclear power offers an important advantage in the case of operations in the Arctic. Conventional submarines are unable to operate under the ice pack because of their inability to ensure they will be able to find ice shallow enough to break through and surface. Without the ability to surface, they will be unable to snorkel and ventilate the boat when their batteries run low and air quality degrades. Nuclear-powered boats have no such concerns. A nuclear ballistic missile submarine (SSBN) can cruise for months under the ice, hiding in a very challenging acoustic environment. This greatly enhances the stealth of these ballistic missile submarines. Only a nuclear-powered attack submarine has the endurance necessary go under the ice and hunt down a SSBN.

In summary, it is stealth that makes the submarine valuable as a military asset. Hiding underwater, it can take advantage of its quiet acoustic signature to operate covertly. It can approach a target unobserved and achieve a military objective. This objective might be a SSK attacking a vulnerable surface ship, a SSN conducting surveillance of a port, or a nuclear guided-missile submarine (SSGN) delivering a special operations team ashore. The SSBN can find an isolated part of the ocean far from adversaries to lie in wait until called to launch its missiles.

The desire to preserve stealth and remain unobserved forces submarines to primarily use passive sensors. Emitting sensor energy would reveal the presence of the boat, robbing the submarine of its most precious tactical advantage. As a result, the submarine's perception of the environment and the tactical situation is more ambiguous and limited compared to what could be achieved with active sensors.

Sensor Capabilities

The need to use only passive sensors imposes serious limitations on a submarine. By its nature, passive sonar is ambiguous. While active sonar provides a bearing and range to a target, passive sonar provides a bearing alone. This requires a submarine to conduct TMA to estimate the position and motion of ships and other submarines nearby. This forces a submarine commander to be more methodical in sensing the tactical environment nearby and slower in approaching a target.

Passive sensors are also not always sufficient to positively identify a target or meet certain rules of engagement. For example, passive sonar may help a submarine to determine that a warship is operating nearby but may not be enough to suitably identify the class or nationality of the ship. The submarine commander may be obliged to expose a periscope to visually identify the target before firing a torpedo. This increases the chance that the submarine itself will be detected and attacked before it can employ a weapon.

The nature of submarine sensors also limits the ability of a boat's crew to perceive the tactical environment. First, submarines generally rely on sensors located on or very near their hull to observe the environment. Sonar arrays are located on a boat's hull or at most several hundred yards astern on a towed cable. The submarine is restricted to observing potential targets from only one point in space, as opposed to an aircraft or ship, which can place sensors over a large area.[2]

The motion of the submarine through the water also limits the ability of the boat to observe the area around it. As the speed of the boat increases, so too does the noise generated by machinery and water moving over the hull. Both machinery noise and flow noise over the hull decrease the sensitivity of passive sonar arrays. This in turn reduces the range at which a submarine can detect ships or other submarines. This greatly influences the search rate that a submarine can generate. Although the sweep width of sensitive sonar arrays on a submarine is large, the speed at which the boat can move and still realize this sweep width is limited.[3]

Having to rely on passive sensors and staying at low speeds results in the submarine experiencing a relatively claustrophobic and ambiguous sensory environment. As a result, boats are most effective when provided with reliable and timely cueing information from commanders afloat and ashore. If an ASW force can attack the C2 system that provides a submarine with cueing information, they have put the adversary boat at a distinct disadvantage. If they can go one step further and present a confusing tactical picture, with decoys that appear to be targets or real targets that appear to be another type of ship, they have degraded the submarine even further.

This reliance on external cueing sources grows even greater for submarines equipped with ASCMs. An ASCM-armed submarine can attack a target at ranges far greater than it could with torpedoes alone. However, this means the submarine often finds itself armed with a weapon whose range greatly exceeds the reach of its own sensors. While some noisy surface ships may be detected at very long range under the right acoustic conditions, guided-missile submarines (SSGs/SSGNs) are often highly dependent on non-organic sensors for targeting their ASCMs. When fighting a submarine force equipped with SSGs and SSGNs, it becomes even more critical to systematically target an adversary's sensor network and C2 system.[4]

Mobility

The mobility of a submarine has always played a key role in how these vessels are employed operationally and tactically. As technology has changed, so too has the speed and maneuverability of submarines in relation to other types of warships. In all cases, however, mobility has been tied closely to stealth and the ability to employ weapons.

To understand submarine mobility, it is best to break the history of the submarine down into two phases. The first era, before the introduction of the snorkel and nuclear power, found submarines operating primarily on the surface. The second era, marked by what some observers have called the "submerged revolution," found submarines sailing primarily underwater, with only brief excursions to the surface to snorkel, receive communications, make an observation via a periscope, or launch weapons.

Before the submerged revolution, submarines were akin to fast torpedo boats that could submerge for short periods. Despite popular misconceptions, early submarines spent the vast majority of their time on the surface. These boats were designed for high speed and good sea keeping on the surface rather than submerged performance. Once underwater, these boats were slow and suffered from limited maneuverability. Their batteries limited them to speeds less than 10 knots and they could only transit underwater for a few dozen miles before they were forced to surface and recharge.

On the surface, early submarines were generally fast enough to run down merchant ships or keep up with some warships. Underwater, however, they were far too slow to keep pace with a surface ship. This imbalance in speed meant that a submarine, once forced underwater, was robbed of a great deal of its mobility and tactical effectiveness.

Operating underwater with reduced mobility had an enormous effect on the tactical employment of a submarine. Early torpedoes were short ranged, relatively slow, and unguided. They were best employed forward of the beam or off the beam of a target. But because a submarine was far slower submerged than its targets on the surface, attacks were generally only possible if the submarine was already located out in front of a surface ship or convoy.

This imbalance between the submerged speed of a submarine and the speed of surface ships gives rise to a relationship known as the limiting line of approach (LLOA). Taking into account the range of a torpedo and the ratio of the submerged speed of the submarine and the speed of a surface target, we can define an area in front of the target where attacks are possible. If a submarine is inside the LLOA, it can close with the target and fire a torpedo. If the boat is outside the LLOA, it cannot get close enough to the ship for a torpedo shot to be effective. From a defender's perspective, a submarine equipped with only torpedoes that is outside the LLOA is no threat at all and doesn't need to be defended against.[5]

A submerged submarine is greatly limited in when and where it can attack targets based on the LLOA concept. But the submarine is even further limited when we take into account its ability to search for and locate targets underwater. Early submarines detected their targets primarily through visual and radar search while operating on the surface. When a ship or convoy was detected, the submarine commander would make use of the high surfaced speed of the boat to race ahead of the target while remaining just over the horizon. Once he was confident his boat was ahead of the target, inside the

LLOA, he would submerge, lying in wait until the ship approached into torpedo range.

The amount of ocean that a submarine could cover with visual or radar search is proportional to the height of the sensor above the ocean surface. A lookout in a surfaced submarine might be as high as 30 feet above the water, and able to scan a sweep width of well over 10 miles. When combined with the high surfaced speed of the boat, early submarines could generate respectable search rates when surfaced.

A submarine moving underwater and searching with its periscope was a completely different matter. Protruding only several feet above the waves, a periscope could scan only a few miles, and had a very narrow field of view. Both these factors drove the sweep width down. The submarine moved far slower below the surface, further reducing the search rate. A submerged boat was limited not only in its ability to attack, but also in its ability to generate high search rates and locate targets.

A submarine on the surface was vulnerable to visual detection and attack by a better-armed adversary. Once spotted, a submarine was usually forced to dive and evade, restoring its stealth but also giving away its mobility. An ASW force that could drive adversary submarines underwater robbed these boats of a great deal of their tactical effectiveness. Once submerged, these submarines were robbed of high search rates and limited in their ability to attack the targets they did detect.

The shift to operating primarily underwater occurred during the 1950s, as conventionally-powered submarines uniformly adopted snorkels and nuclear-powered submarines began to deploy. To counter submarines that operated out of the reach of visual and radar search, navies turned to passive and active sonar. Sonar array design and acoustic signal processing techniques advanced rapidly as ASW forces struggled to hunt underwater.

Just as sensor technology advanced, so did propulsion technology and naval architecture. Rather than shaping the hull of a submarine to operate on the surface, naval architects began to build streamlined submarines that were optimized for underwater speed and maneuverability. Conventional submarines were faster underwater than on the surface. Nuclear reactors generated so much energy that SSNs were able to match or exceed the speed of surface ships. These advances had important implications for how submarines were employed tactically.

The ability to move quickly underwater imparted advantages and disadvantages in these new submarines. Speed allowed a submarine to reposition rapidly, quickly evade attack, and approach a surface ship from the quarter or astern. But moving at high speed generated a great deal of noise, which could be readily detected by ASW forces with passive sonar. At high speed, water flowing over the hull generated large amounts of low-frequency noise. Running at high speed caused nuclear submarines to generate a great deal of noise, since they were obliged to run their coolant pumps in noisy, high-capacity modes to move coolant from the reactor to the rest of the propulsion plant. Even the quieter conventional submarines became much more vulnerable at high speeds. To move faster, the screw of a submarine must spin at a higher speed, which can cause a phenomenon known as cavitation. Cavitation occurs when a fast-moving screw causes bubbles to form that quickly implode and generate loud noises as the seawater rushes to fill the void. This phenomenon impacted both conventionally-powered and nuclear-powered boats.

When discussing modern submarines, it is useful to talk about a maximum underwater speed and a slower speed, below which the boat remains relatively quiet and hard to detect. This slower speed is often known as "maximum tactical speed." Below maximum tactical speed, flow noise over the hull is minimal and cavitation is unlikely to occur. A submarine commander attempting to remain covert is unlikely to operate above maximum tactical speed unless there is a pressing need to do so.

At first, it might appear that a submarine that is faster than a surface ship does not suffer from LLOA. From a strictly kinematic perspective, this is true. A modern SSN can easily close with an average merchant ship from any angle and get close enough to launch a torpedo. However, chasing a target from astern forces the submarine to move at high speeds, where it is readily detectable and very vulnerable to ASW forces equipped with passive sonar. Again, we see the close link between mobility and stealth.

When analyzing the maximum tactical speeds of modern SSNs and the maximum speed of a likely target, a modern merchant ship, we find that a very interesting relationship occurs. While the maximum tactical speed of submarines has increased over the past seven decades, so too has the maximum speed of most merchant ships. Despite major improvements in naval architecture over the past 70 years, average LLOA have remained nearly constant. If a submarine commander wishes to remain stealthy and operate below maximum tactical speed while attacking, he is still obliged to attack with torpedoes from a sector forward of the target.[6]

The ability to operate at high speeds gives nuclear-powered submarines an important advantage over conventional submarines. An SSN or SSGN can sustain very high speeds for days at a time, allowing it to rapidly transit over hundreds or even thousands of miles. This provides both the ability to sortie rapidly to a patrol area and the endurance to hide or hunt there for many weeks at a time. By their nature, conventional submarines are far more kinematically limited. While they may have the fuel to transit long distances, they cannot do so quickly. They must frequently snorkel and expose themselves to being detected. This has led some observers to refer to conventional boats as functioning more as "mobile minefields" than as true submarines.[7]

A nuclear submarine retains the ability to close with or evade targets at high speed, even if this makes the boat more vulnerable to detection by passive sonar. This allows a SSN to have much more control over a tactical scenario. A conventional submarine, however, is much more limited. It can only sustain high speed for at most several hours at a time. Unless a torpedo-equipped conventional submarine sprints at high speed, and in doing so exhausts its batteries, it is forced to lie in wait ahead of surface ships. This reduces the likelihood of a successful attack by a conventional submarine to a matter of chance that it is located in the pathway of an approaching warship or merchant vessel. Conceptually, this is similar to the problem of a ship traversing a minefield, where it has a particular change of triggering a mine depending on the geometry of the minefield. Indeed, it seems the description of conventionally-powered submarines as mobile minefields is a useful analogy.[8]

Endurance

On the tactical level, the powerplant, environmental systems, and speed at which the boat is operated will affect how often the crew must conduct evolutions that make them more readily detectable. On the operational level, factors such as crew fatigue, supplies, fuel, and weapons expenditure will drive how long a boat can remain on patrol.

When discussing tactical employment, our primary focus will be on how long the boat can preserve the dual advantages of mobility and stealth. Every submarine, regardless of its type of powerplant, has to conduct housekeeping functions that are noisy and degrade stealth. Compressors must be operated to replenish high-pressure air tanks, trash must be dumped overboard, and pumps must be run. Evolutions like these are often noisy, and wise submarine commanders will generally try to find areas of the ocean where they can blend in while they conduct these tasks.

On top of noisy housekeeping, conventional submarines must periodically surface or expose a snorkel to run their diesel engines, recharge their batteries, and refresh the atmosphere inside the hull. This makes the boat far more vulnerable to detection by radar and to a lesser degree passive sonar, since diesel engines are far noisier than an electric motor running off a battery bank. The length of time that a conventional submarine can go without snorkeling is directly tied to how rapidly the batteries are drained of power.

Charged batteries provide commanders of conventional submarines with mobility. They give them the speed and endurance to close with a target or the maneuverability to evade detection and attack. Preserving high levels of battery charge is therefore a centerpiece of SSK tactics. A conventional submarine can achieve relatively high speeds underwater, but cannot do so for long, since this rapidly drains the batteries. The slower a SSK patrols, the longer it can remain submerged before snorkeling.

Nuclear submarines have no such limits on their underwater mobility nor do they need to surface. Their reactors require no outside air to operate, and the large amounts of power they generate allow them to scrub carbon dioxide from the air. This removes any requirement to surface and attend to the powerplant or environmental systems. An SSN is far less vulnerable to detection by radar because it has little need to surface for extended periods. A nuclear submarine commander may still expose a periscope or sensor mast for tactical reasons or a communications antenna to send or receive messages, but these excursions are very rapid and expose the boat far less than do the prolonged indiscretions of conventional submarines. Compared to conventional boats, nuclear submarines have essentially unlimited endurance in terms of tactical mobility.

Submarines equipped with AIP systems are an interesting hybrid, with characteristics of both conventional and nuclear submarines. Their AIP equipment allows them to stay submerged for many days or even weeks at a time. Additionally, they can sustain slightly higher speeds for longer periods, increasing their mobility compared to a conventional submarine. The true strength of an AIP submarine, however, lies in its ability to reduce its indiscretion rate.

On an operational level, fuel, crew endurance, supplies, and weapons expenditure drive how long a boat can remain at sea. Nuclear-powered boats have essentially unlimited range, as their reactors can power the boat for at least a decade without refueling. Conventional and AIP-equipped submarines, however, burn through their fuel as the patrol goes by.

Cramming a powerplant, environmental systems, weapons, and a crew into a small but seaworthy vessel has always been a challenge for naval architects. Often times, it was the quality of crew accommodation that was sacrificed to make room for other systems. Historically, this has been truer aboard conventional boats, which suffered from cramped berthing, poor environmental systems, fumes from diesel engines, and rough sea keeping during surfaced operations or snorkeling. The ability of a crew to endure these conditions can drive how long a boat can remain at sea.

In the case of nuclear submarines, the living accommodation is far more comfortable. To make room for the nuclear plant, SSNs and SSGNs are often far larger than their more diminutive conventional cousins, leaving more room for crew berthing. Large amounts of power allow for more advanced environmental systems, making the boat more comfortable. Nuclear-powered boats operate deep, largely free from the sickening motion caused by waves on the surface or at periscope depth. What does constrain these boats, however, is the amount of food and supplies they carry. Frequently, it is the stock of provisions that limit the patrol length of a nuclear-powered submarine.

During wartime, the stock of weapons aboard influences the length of a submarine patrol. The historical record during both World Wars frequently found attack submarines exhausting their torpedoes before they consumed all the food or fuel aboard. Depletion of magazines is often exacerbated by the poor reliability of torpedoes during combat. Torpedoes are complex weapons, which must endure months or years at sea under challenging conditions yet operate flawlessly when fired.

It is very difficult and expensive to simulate wartime tactical conditions during peacetime weapons testing. Naval bureaucrats and arms manufacturers also have little incentive to stress their weapons and expose them to failure before a conflict. Poor wartime torpedo reliability has been experienced by several different navies in various conflicts during the last century. It is possible these challenges will persist in the future, making the rate of weapons expenditure a driver of the endurance of a submarine during a combat patrol.[9]

Weapons

Understanding the types of weapons carried by a submarine as well as their advantages and limitations helps aviators to defend against their effects. The guidance and kinematic capabilities of submarine-launched weapons drive the tactics used by submarine crews during an attack.

The primary weapon employed by submarines is the HWT. The speed and range of these weapons are good, but it is the large warhead carried by the torpedo that makes it a ship-killing weapon. The keel of a ship is not armored, making it very vulnerable. Detonating a warhead under the keel creates an air bubble, which travels upward, lifting the ship as it does, before collapsing. As the bubble collapses, the hull of the ship sags downward, severely stressing and possibly breaking the keel of the ship as it does so.

The simplest HWTs are straight-running weapons controlled by a gyroscope. They are set to run at a particular heading and generally fired in a spread towards a target ship. Modern HWTs feature seeker heads to help guide them to a target. These seeker heads are equipped with passive sonar, active sonar, or wake detection sensors.

To carry out a successful attack, the torpedo must be able to travel the distance from the submarine to the target ship. After closing the distance, the torpedo must pass close enough to the target for its fuse to function or for the seeker head to detect the target and begin to guide the weapon to impact. The seeker head's field of view is limited, meaning that it is critical a submarine crew must deliver the torpedo to a point in space where it stands a good chance of the target being nearby.

This is not as simple a solution as it might seem. The quality of the sensor information used to generate a firing solution may be poor, especially in a high-threat environment. A warship alerted to an incoming torpedo will likely evade aggressively, reducing the chance that the ship will be within range of the seeker head when the torpedo arrives nearby.

To address this problem, torpedo designers developed two guidance methods to aid a weapon after launch. The first is wire guidance, in which the torpedo spools out a small cable that remains connected to the submarine after launch. Using the wire, the submarine can send updated information to the torpedo as it closes the distance to the target. These wires are relatively fragile, however, which forces the submarine to avoid high speeds or violent maneuvers to prevent breaking the umbilical. Typically, it is not possible to reload a torpedo tube without cutting the wire, reducing the submarine's weapons-launch rate.

An alternative to wire guidance is equipping torpedoes with sensors that can detect the wake of a surface ship. The wake of a ship moving at medium or high speed is slow to dissipate and easily detected underwater. Because the wake is a much larger target than the hull or sound signature of a warship, it is a far easier fire-control problem to deliver a torpedo under a wake than it is under a hull. Once the torpedo has detected a wake, it can begin to turn under the wake and follow the disturbed water towards the ship.

Wake-guided torpedoes are far easier for inexperienced submarine crews to deliver than are more complex wire-guided torpedoes. However, these weapons suffer from range limitations because they must travel to a wake and then expend even more fuel following the wake towards the target. Just as wire guidance constrains how a submarine may maneuver after launch, so too does wake guidance constrain how a torpedo is employed tactically.

Because of the factors listed above, the effective range from which a torpedo is employed is often far less than the actual kinematic range of the weapon itself. Weapons capabilities, seeker sensitivity, crew training, and doctrine will strongly influence the range at which submarine commanders choose to fire their weapons.

Like torpedoes, submarine-launched ASCMs are often employed at ranges far shorter than their maximum kinematic range. Frequently, submarine-launched ASCMs have ranges far longer than the sensor range of the boat that launches them. Just like a homing torpedo, an ASCM must be delivered to a sensor activation point where its seeker head can detect the target ship and guide the rest of the way. Developing a precise-enough track of a ship from long range may be very challenging, or even impossible, for a submarine. This means a boat may be forced to expend many weapons to raise the probability that one of the many seekers acquires. It also means that employing ASCMs may be undesirable if neutral ships are nearby because of fears of collateral damage. As a result, ASCM-armed submarines are highly dependent on off-board cueing and friendly C2 systems to deliver targeting data.

Communications

The opacity of the water hides submarines from non-acoustic detection but also makes communication difficult for a submerged boat. A submarine underwater is in some respects an island unto itself, cut off from communicating with friendly forces. Submarines must make the best of their sensor suites to develop an accurate picture of the tactical situation around them.

A submarine can expose communications masts to send and receive information. However, any transmission a boat makes stands some chance of being intercepted and geo-located. Since the earliest days of ASW, nations have invested in signals-gathering equipment and worked to plot the location of submarines as they transmit to commanders ashore. High frequency direction finding (HF/DF) and geo-location by SIGINT equipment give ASW forces cueing data that is helpful in hunting down a boat on patrol. As a result, submarine crews will do their best to limit the frequency and duration of transmissions off the boat.

Submarines underwater do retain the capability to receive messages from commanders ashore. Extremely low frequency antennae can transmit signals which penetrate deep underwater. While their data transmission rate is extremely slow, they are useful to call a submarine to periscope depth, where it can expose a communications mast and receive message traffic from a satellite. Alternatively, some submarines are equipped with floating communications buoys, which can be trailed on the surface behind a submerged boat to copy down line-of-sight transmissions or satellite communications broadcasts.

Self-defense

The ability of a submarine to defend itself from attack has varied greatly over the last century. What has remained constant, however, is that submarines are relatively lightly armed and incapable of absorbing heavy damage. The underwater environment is very unforgiving, making damage to a boat and flooding much more threatening for a submarine than for a surface ship. Submarines have always had to respect better-armed ships and aircraft. With certain exceptions, their light construction means they cannot withstand torpedo or close-aboard depth-charge attack. They must rely on their stealth and mobility to keep them safe, rather than relying on robust self-defense.

The earliest submarines were frequently armed with deck guns for surface engagements. However, their light construction and lack of armor meant that their crews would rather submerge than stand and fight against a better-armed enemy. Even token resistance from armed merchant ships could be sufficient to drive off a submarine, as the crew would be unwilling to risk hull damage that could prevent them from submerging and leave them vulnerable on the surface.

Early submarines also featured AAA to drive off attacking aircraft. During World War I, the very slow speed of ASW aircraft meant that these weapons were infrequently used. A submarine could usually quickly dive and evade long before the primitive aircraft arrived within bombing range. Still, a nearby aircraft could alert other ASW forces to the presence of a submarine, and this robbed a submarine of its stealth. Often, the presence of aircraft was enough to force submarines to submerge, not so much because the crews were afraid of the lightly armed and slow airplanes, but because staying surfaced made them conspicuous to patrolling warships.

As aircraft became faster in the interwar period, they became much more serious threats to submarines. Generally, an alert lookout aboard a submarine could detect an approaching aircraft before the aviators could spot the submarine. This usually gave the crew enough time to submerge before the aircraft could close the distance and deliver an attack. If the submarine sighted the aircraft and submerged late, however, they put themselves in deadly peril. Once the crew went below and abandoned their AAA, the aircraft had an unopposed run over the exposed hull of the partially awash or just submerged boat.

As a result, submarine anti-aircraft tactics in World War II hinged around how far away an aircraft was detected from the boat. If the submarine could submerge fully before the aircraft could close the distance, the crew would rush below and carry out a crash dive. If the aircraft were sighted close to the submarine, the crew would stand and fight on the surface, hoping to damage or shoot down the incoming aircraft as it closed to attack.[10]

After the submerged revolution, deck guns were made obsolete and removed from submarines worldwide. Operating underwater, the submarines had no realistic way to track an aircraft overhead and fire a weapon to drive it off. To attack the boats underwater, aircraft equipped themselves with depth charges and torpedoes. To counter these weapons, submarines relied on evasion and countermeasures. Evasive maneuvering could throw off the ability of an aircrew to precisely locate the submarine. Once a torpedo was in the water, the submarine could deploy acoustic countermeasures to confuse the guidance system of the incoming weapon.

After the submerged revolution, the light hull and inability of boats to attack airborne threats meant that submariners were obliged to take hostile ASW aircraft and their weapons seriously. They could not afford to ignore the threat of an inbound torpedo. This fact drives modern submarines to be very sensitive to any potential threat platform or weapon. A submarine that detects a torpedo in the water nearby is obliged to evade aggressively, because its survival depends on eluding the weapon.

In discussing submarine vulnerability, it is worthwhile to discuss the design philosophy of the Soviet Navy and the construction of several types of boats by Soviet design bureaus. During most of the Cold War, the United States and its NATO allies possessed submarines with a large measure of acoustic advantage. Western boats were quieter than their Soviet adversaries and equipped with much more sensitive sonar suites. As a result, the Soviets expected that they themselves would be detected by Western boats long before they were aware of the presence of an interloper. The Soviets expected Western submarines to frequently fire first and equipped their boats with high-speed 'revenge' weapons and robust hulls that offered the best change of absorbing a torpedo hit and continuing to fight.[11]

Many Soviet and current Russian Federation boats are built with double hulls. The internal pressure hull surrounds the habitable compartments. An external hydrodynamically-shaped hull provides an additional layer of structure that a torpedo warhead must penetrate in order to cause damage or flooding. These boats have high levels of reserve buoyancy to deal with flooding and are often planned to be "unsinkable" when on the surface.[12] As a result, submarines built by the former Soviet Union and Russian Federation proved very challenging targets for Western ASW aircrews through the later years of the Cold War and in the modern day. The adoption of lightweight torpedoes (LWTs) with penetrating shaped-charge warheads are a testament to the design philosophy and hardening of Soviet submarines.

Notes

1. Certain nuclear-powered submarines have been designed in such a manner to use variations in temperature and the resulting convection currents to move pressurized water through the core at low power levels. This method is called natural circulation and has been adopted in several American submarine classes. Still, at high power levels, the amount of heat generated in the core obligates nuclear submarine crews to operate their coolant pumps. Magdi Ragheb, "Naval Nuclear Propulsion," in "Nuclear Power – Deployment, Operations and Sustainability" (London: InTech Open, 2011), pp. 2–30.
2. For example, a patrol aircraft can drop sonobuoys spaced dozens of miles apart. An air-capable warship can launch a helicopter or unmanned aerial vehicle that can search many miles from the ship. This ability to spread sensors over wide areas away from own ship draws a distinction between aircraft and warships and the more localized sensing ability of a submarine. This relationship may change in the years ahead as submarines begin to employ unmanned underwater vehicles.
3. The American *Seawolf* class SSN is the prime example of a submarine design optimized to generate high search rates. In the 1980s, the U.S. Navy began to pursue a strategy that centered around hunting down Soviet SSBNs in "bastions" under the Arctic ice, where they were protected by cordons of surface ships, aircraft, and SSNs. If American SSNs could penetrate the bastions and hunt down the SSBNs during the opening days of a conventional conflict, it would inject uncertainty into the minds of Soviet leaders. It would also force them to pull their most capable SSNs out of the supply lanes in the North Atlantic and back under the ice to protect the SSBNs. This virtual attrition would make it far easier for NATO to maintain supply lines across the North Atlantic and resupply beleaguered forces fighting a conventional conflict in Western Europe. In order to hunt down the Soviet boomers, the U.S. Navy required a boat that could operate independently of external cueing. It had to search large areas rapidly with organic sensors. It needed a large weapons payload to fight its way through cordons of warships and SSNs and continue hunting. The result was a boat with advanced sensors, excellent quieting, high tactical search speed, and a very large weapons load. These allowed a *Seawolf* to generate high search rates, and continue killing on-station without having to withdraw to rearm. Construction of this class of submarines was truncated after the collapse of the Soviet Union due to high costs and significantly reduced threat, not lack of capability.
4. There are several historical and current examples of sensor networks and C2 systems supporting submarine operations. For example, during World War II German U-boats were directed from headquarters to form wolfpacks in particular positions and target convoys whose position was known from scouting or cryptanalysis. During the Cold War, American SSNs were guided to areas to search for targets by SOSUS underwater sonar arrays. In a similar manner, Soviet SSGNs trained for anti-shipping operations

where they would be supplied with targeting information by radar satellites, electronic intelligence satellites, or reconnaissance aircraft. A modern example is Chinese ASCM-armed submarines supported by land-based over-the-horizon radar.

5. For an in-depth discussion of the LLOA concept, see Bernard Osgood Koopman, *Search and Screening: General Principles with Historical Applications* (Alexandria, VA: Military Operations Research Society, 1999), pp. 33–6.

6. Bryan Clark and John Stillion, "What it Takes to Win: Succeeding in 21st Century Battle Network Competitions" (Washington, D.C.: Center for Strategic and Budgetary Analysis, January 22, 2015), pp. 26–7.

7. The geography and constricted waters near the Soviet Union shaped operational environments very conducive for employing conventional "diesel" submarines. Well aware of the low radiated noise levels but limited mobility of these diesel boats, the Soviets referred to these submarines in their own literature as "mobile minefields." Consider the following remarks by Captain Thomas A. Brooks, USN (ret.), former Director of U.S. Naval Intelligence from July 1988 to August 1991: "The Soviets see a continuing utility of the diesel submarine. It is excellent for confined waters in the Mediterranean; it makes a superb 'mobile minefield' in Soviet parlance." Thomas A. Brooks, "(Soviet) Diesel Boats Forever," U.S. Naval Institute *Proceedings* (December 1980), p. 107.

8. Andrew Erickson, Lyle Goldstein, William Murray and Andrew Wilson, *China's Future Nuclear Submarine Force* (Annapolis, MD: Naval Institute Press, 2007), p. 67.

9. The poor reliability of German and American torpedoes during World War II is well documented, but concerns about torpedo reliability are not only restricted to this period. The last conflict to involve a Western submarine sinking a surface ship found the submariners employing an ageing but reliable torpedo rather than a more modern and sophisticated weapon. During the campaign to recapture the Falkland Islands in 1982, HMS *Conqueror* (S48) sank the Argentinian cruiser ARA *General Belgrano* with ageing and unguided, but highly reliable, Mark VIII torpedoes rather than the modern Spearfish torpedo. The Mark VIII was first introduced before World War II, although it was much modified between that time and its eventual retirement in the early 1990s.

10. Alfred Price, *Aircraft Versus Submarine in Two World Wars* (Barnsley, South Yorkshire: Pen and Sword Books, 2004), pp. 11–12 and 148–9.

11. Norman Polmar and K. J. Moore, *Cold War Submarines: The Design and Construction of U.S. and Soviet Submarines* (Washington, D.C.: Potomac Books, 2004), pp. 303–04.

12. These submarines are not unsinkable in the literal sense of the word, but rather possess a sufficient level of reserve buoyancy that several compartments can be flooded but the boat still remain safely afloat.

Chapter 4

Airborne ASW History: Genesis to World War II

S ubmarines and aircraft traced parallel development paths during the last century. Both were adopted for military use in the prelude to World War I and came of age during that conflict. During World War II, submarines were used by both the Allied and Axis powers to carry out highly effective anti-surface warfare (ASUW) campaigns. Allied aircraft helped break the back of the German U-boat fleet during the Battle of the Atlantic. With the dawn of the Cold War and the introduction of nuclear weapons, submarines became strategic weapons platforms. ASW aircraft were called on to serve as a first line of defense, flying missions to locate these boats and hold them at risk. Modern submarines are impressive multi-mission platforms, capable of gathering intelligence and striking targets in multiple domains. Modern patrol aircraft play a key role in defending against them.

In these next two chapters, we will examine some broad trends to better understand how submarines and aircraft have historically been employed. We will review how technical advances drove improvements in sensors, weapons, and the capabilities of the platforms themselves. We will try to understand how strategic challenges shaped the force structure of various air arms and submarine fleets. Our review will be broad and general in nature rather than specific and detailed. The goal will be for aviators and planners to have a basic understanding of the trends that drove aircraft and submarine development.

The Beginning

In the mid-nineteenth century, the British Royal Navy enjoyed an unchallenged reign over the world's oceans. Conscious of their inability to win a force-on-force conflict, various nations sought to build weapons that could undermine the Royal Navy in an asymmetrical manner. By the 1860s, European and American shipbuilders had begun to experiment with undersea vessels. Eager to undermine the British, the French invested in a fleet of torpedo boats and early submersibles. During the late 1890s, an American inventor named John Holland built a series of submersibles that ushered in the era of the modern submarine.

Holland's designs and others like them were more submersible torpedo boats than true submarines. They spent the majority of their time on the surface, diving only to attack or to evade more powerful adversaries. Once submerged, these boats were slow, un-maneuverable, and very limited in their range.

Still, these early submarines allowed nations to extend the reach of their coastal defenses beyond the range of land-based artillery. While unable to keep up with a line of fast cruisers or battleships, they were able to act as scouts for the larger warships. In addition, they were inexpensive to build and operate. These early boats were equipped with gasoline engines to power them while surfaced and battery banks and electric motors to drive them while submerged. Gasoline engines were inefficient and prone to accidents, limiting the cruising range of these early submarines and making them dangerous for their crews. While a primitive start, these early submarines opened the door to rapid advances in the first decade of the twentieth century.

The Wright brothers took to the air in 1903. Their aircraft and others that followed were very immature. Their controls were primitive and poorly developed, making these early aircraft demanding and dangerous to fly. Their engines were underpowered and unreliable. These first-generation aircraft were extremely limited in their speed, range, and load-carrying capabilities.

However, aviation technology did not stand still. In the ensuing decade between the Wright brothers' first flights and the outbreak of World War I, aircraft design advanced rapidly. Fuselages and wings became more robust. More powerful engines allowed aircraft to reach higher altitudes and higher speeds. Additional power also brought greater load-carrying capability.

On the eve of World War I, many officers and military thinkers were aware of the general capabilities of aircraft and submarines. However, most believed that both platforms would remain interesting toys of little practical military value. What was not appreciated at the highest levels of leadership was how rapidly these two platforms had matured in the years leading up to the conflict.[1] Submarines had traded their gasoline engines for diesels, improving their range and safety for their crews. Submarine designs had shifted from circular hulls optimized for submerged operations to wedge-shaped double hulls optimized for surfaced speed and seaworthiness. While many submarines were small and intended primarily for coastal defense, several navies had built ocean-going submarines armed with powerful torpedoes and capable of ranges in excess of 5,000 nautical miles.

Both the Entente and the Central Powers entered the war with little consideration of the role submarines might play in commerce raiding. This was due to the particulars of international law. A warship was legally permitted to sink enemy merchant ships, but only after merchant mariners and passengers had been removed from the civilian vessel. Neutral merchant vessels could not be sunk at all. They could be stopped and inspected for contraband, but only kept as prizes if illicit cargo were found. Due to the small size and limited crews of submarines, it was assumed taking passengers onboard or manning captured merchants with prize crews would both be impossible. As such, there was little worry about submarines menacing merchant fleets.

Leaders in the Entente believed it highly unlikely that another nation would violate international law and attack merchant ships. As a result, the Royal Navy paid little attention to the importance of defending merchant vessels. Instead, the service focused on its traditional role of bringing the enemy fleet to a decisive battle and thereby establishing sea control.[2] This single-minded focus on offensive action had great influence on the approach the Royal Navy took to meeting the submarine threat during the ensuing years of the conflict.

Some naval officers on both sides had considered the threat that submarines posed to warships. The limited speed of submarines and their vulnerability on the surface meant they were unlikely or unable to pursue and attack the battle line. Similarly, their low underwater speed and endurance meant these boats would be hard pressed to catch and torpedo a fast warship while submerged. Submarines could best threaten powerful surface ships by scouting for the opposing fleet or lying in wait in hopes that a battleship or cruiser might pass overhead and make itself vulnerable for a well-timed torpedo attack.

Before the outbreak of the war, a small numbers of officers in the United Kingdom and Germany had considering the problem of actually fighting submarines, and how aircraft might be used. Between 1911 and 1912, the British and German militaries each carried out a series of trials to determine whether aircraft were capable of visually detecting a submarine.[3] The trials were dangerous, since the poor reliability of early airplane engines made extended flights over water a very risky affair. However, it was clear that aircraft could detect a surfaced submarine, and in theory, were capable of attacking the boat, provided they were equipped with suitable weapons.

World War I

Fighting between the Entente and the Central Powers began in August 1914, and by the end of the year, submarines and aircraft had already proved themselves in battle. It became readily apparent that both platforms were not toys, but rather serious weapons systems in their own right. Initially pressed into service as aerial observers, aviators on both sides quickly took to arming their aircraft with guns and improvised bombs. German submarines scored several stunning victories over the Royal Navy, sinking powerful warships and leaving the British reeling.[4] The opening months of the war also witnessed the first attack by an aircraft on a submarine, when a German airship sighted and bombed a British submarine.[5]

By 1915, with any hope for a rapid end to hostilities gone, the British instituted a full blockade of Germany. The disruption of supplies and trade led to economic turmoil and food shortages. By February, widespread public outcry led the Germans to retaliate with submarine attacks against Entente merchant ships.[6] The waters around the British Isles were declared a "war zone," where enemy ships and neutrals should "navigate at their peril."[7]

The small numbers of operational U-boats limited the initial losses of merchant ships. So too did German rules of engagement, which required U-boat captains to close to short range, surface, and identify the nationality of their target prior to an attack. Each of these steps made the submarines vulnerable to counterattack. Worldwide outcry against attacks on merchants led the Germans to curtail their submarine warfare campaign in September 1915.

Few Entente leaders were much concerned after the end of the first German submarine offensive. The small number of merchant ships sunk made the threat of future submarine campaigns seem minimal. In reality, however, the low number of sinkings was due almost entirely to the small number of German U-boats in service and their self-imposed rules of engagement rather than limited lethality of the submarines themselves.[8] Thus, the nearly nonexistent ASW capabilities of the Royal Navy and the true nature of the threat posed by the U-boat fleet both went unacknowledged.[9]

In reality, British warships possessed almost no ability to detect or damage a U-boat operating underwater. Thus, as the war dragged on, the Royal Navy was forced to adopt ASW methods to degrade the U-boats in a passive manner. The submarine's slow submerged speed meant it couldn't keep up with surface vessels while submerged. The tiny field of view and limited height of eye of the periscope meant boats could not adequately search for targets either while underwater. Forcing U-boats to submerge either with surface ships or aircraft therefore became a very useful tactic.

The British employed minefields and barriers to degrade the U-boat fleet. A series of nets and buoys known as the "Dover Barrage" was deployed across the English Channel. This was intended to either ensnare transiting U-boats or force them deeper where they would set off mines moored to the ocean bottom. While the barrier was difficult to maintain and led to only a few U-boats being sunk, it led many German submarine commanders to travel the longer route around Scotland and Ireland to reach operating areas in the Western Approaches. A longer time spent in transit meant the U-boats spent less time hunting and sinking during a patrol. This case of virtual attrition caused by the Dover Barrage was unintentional on the part of the British but posed an operational penalty on the Germans.

Entente aircraft were also tasked to fly patrols to locate U-boats and force them to submerge. Because the chances of finding a submarine in the open ocean were small, the aircrews focused on known U-boat transit lanes.[10] The also British flew unarmed trainer aircraft known as "Scarecrows" on anti-submarine patrols in coastal waters. These aircraft couldn't attack if they spotted a U-boat, but they could force the boat to submerge and report its position to warships nearby.

In 1916, the Germans resumed submarine attacks on merchant ships. The U-boat crews focused their efforts on the waters where merchant ships tended to funnel together as they left the open ocean and began their approach to port. Here, the submariners could find the highest density of targets. German commanders chose operating areas far enough from shore to evade most of the short-range airborne and surface patrols.

Despite increasing losses, British leaders stubbornly refused to adopt convoy tactics to protect merchant ships. Commanders in the Royal Navy and merchant mariners generally believed it was impractical to operate a large number of merchant ships together. Herding merchant vessels was a dull and distinctly defensive task, one that the offensive-minded British officers were loath to undertake. Additionally, the British believed that a submarine on patrol was less likely to detect several dozen ships sailing alone than one large group of ships sailing together.

On this later fact, the Royal Navy was badly mistaken. Statistically, a submarine was much more likely to spot hundreds of vessels sailing by themselves than to sight several large groups of merchant ships sailing together. Therefore, by simply operating groups of ships together, the British could greatly degrade the ability of German U-boats to find targets.

Sailing in a convoy also had several additional operational benefits that tilted the advantage sharply towards the ASW forces. First, convoys acted as force multipliers for the escorting warships. When a U-boat crew detected a convoy, they and their compatriots nearby would do their best to join and attack. Because convoys drew submarines towards them, a warship was far more likely to detect a submarine while escorting a convoy than it was while conducting independent ASW patrols.

Second, convoys sharply increased the lethality of the threat environment for attacking U-boats. A submarine attacking a lone merchant faced no true threat and a crew attacking a lone warship need only evade the defenses of a single vessel. When attacking a convoy, however, U-boats needed to evade the collective defenses of a team of warships working together. Sailing merchants and warships together forced U-boats to fight for their lives every time they approached to attack.[11]

Last, convoys reduced the rate of losses because submarines could not usually attack more than one ship at a time. By the time a U-boat had approached, attacked, and maneuvered to evade the escorts, the other ships of the convoy had usually sailed out of range.

In February 1917, the German Navy undertook a concerted effort to cut off seaborne trade to the United Kingdom, and in doing so, end the war.[12] German U-boats began unrestricted attacks against merchant ships, racking up staggering numbers of sinkings. At one point, one out of four British merchant ships were being sunk before they completed a round trip voyage.[13] The British faced the distinct possibility that they would be forced to sue for peace.

By May of that year, the Royal Navy reluctantly began to organize convoys. Ships sailing together had an immediate and drastic effect on shipping losses. The British took additional advantage of SIGINT while planning convoys. German U-boats communicated to their commanders ashore using high-frequency radios. The British were able to plot these transmissions and route convoys around areas of expected submarine activity.

The effectiveness of the British convoy system forced U-boats to alter their tactics. Instead of hunting for targets in the approach zones, the submarines focused on attacking targets after the convoys had broken up. The U-boats moved to coastal waters, where they stalked merchant ships in the final miles from the convoy dispersal areas to ports of arrival. By the final months of the war, two-thirds of sunken merchant ships had been attacked within 10 miles of land.[14] In these new U-boat operating areas, the British flew frequent "Scarecrow" patrols, trying to locate U-boats, force them to submerge, and rob them of the benefits of mobility and the element of surprise.[15]

In the final months of the conflict, the Entente powers were experimenting with ways to detect a submerged submarine. The British conducted experiments using hydrophones lowered underwater by seaplanes that had landed on the surface. Airships were also used as hydrophone platforms. While these methods were crude and generally ineffective, they marked an important milestone in the effort to strip the submarine of its stealth.

By the end of World War I, several themes stood out clearly. Both aircraft and submarines had proven themselves as effective weapons platforms. Submarines were very effective at ASUW, both in the open ocean and in the littorals. Patrol aircraft were very limited in their ability to damage surfaced submarines because of their slow speed and limited firepower. They had no ability to detect or attack submerged boats. However, they had proven very effective ASW platforms because they could degrade the effectiveness of submarines through passive means such as forcing submarines to submerge.

Passive means of ASW had proven to be key strategies during the war. Convoys led to large reductions in losses of merchant ships. Minefields and air patrols had caused virtual attrition against U-boats. By the time the Armistice was signed by the belligerents, the first foundations of ASW had been laid.

The Interwar Period

During the two decades between the defeat of the Central Powers and the beginning of World War II, important advances were made in ASW equipment and tactics. Physicists and engineers experimented with radio waves and sound energy to detect submarines on the surface or underwater. British and American engineers developed rudimentary equipment that could emit sound energy, which in turn would reflect off submerged boats and return to a receiver. Called Sound Navigation and Ranging (SONAR) in the United States and Asdic in the United Kingdom, this technology was the forerunner of the active sonar systems used today.

During the interwar period, airpower theorists expounded on the reach and power of the heavy bomber. Rapid improvements in powerplants and aerodynamics had resulted in vast improvements in the speed, range, and load-carrying capabilities of large multi-engine aircraft. Many of these bombers were faster than the fighter aircraft of the day, leading to claims that absent some wonder weapon, defense against a modern bomber force would be an exercise in futility. Airpower enthusiasts boasted that armed with modern bombers, an air force could pulverize the industrial resources and cities of an enemy, bringing an end to a conflict through airpower alone.

Improvements in large aircraft had three important effects on ASW aircraft. The first was the improved load-carrying capacity of multi-engine aircraft. Unlike the fragile ASW aircraft of World War I, patrol aircraft in the interwar period carried enough fuel to provide very long range and excellent endurance. They had high payload capacity, allowing them to carry numerous powerful ASW weapons. They were large enough to carry a dozen crewmembers to operate sensors and act as lookouts. Their much higher top speeds compared to World War I patrol aircraft allowed these new aircraft to rapidly close with and attack a submarine before it could submerge.

The increased potency of bomber fleets drove a search for novel approaches to air defense. British military leaders and politicians were especially concerned about the threat of bombers attacking from the Continent. The close proximity of airfields in France and Belgium to cities in England meant that the RAF would have precious little time to launch and intercept enemy bombers before they reached their targets. Military leaders hoped in vain for a "death ray" that could blast bombers out of the sky. While this technology was a fantasy, investigations into electromagnetism led to the development of radar.

Early radar sets were not capable of being carried onboard aircraft during most of the interwar period due to technical limitations. The British possessed a radar set that was sufficiently precise to detect a surfaced submarine or other small ship but they lacked power-generating equipment light enough and compact enough to fly aboard an aircraft. Prior to the outbreak of hostilities, such a power source was not available in the United Kingdom, the United States, or Germany. This limitation delayed the introduction of airborne ASW radars.

In the lead-up to World War II, air arms began to improve the designs of ASW weapons. As early as 1924, the British had begun developing specialized bombs for attacking submarines. These weapons had fuses that could detonate after a brief delay underwater or if they struck the hull of a submarine.[16]

The Allied and Axis powers entered the conflict in 1939 with great belief in the coercive power of strategic bombing. Despite the failure to develop a fleet of heavy bombers, the Nazis had great faith in the ability of the Luftwaffe to smash British resistance through a strategic bombing campaign. The British viewed RAF Bomber Command as the key tool to strike back against Germany with powerful reprisals should the war continue. Both the Allies and Axis marched to war with much more attention focused on their fleets of fighters and bombers than on their submarines and ASW aircraft. This misplaced focus on the aerial threat instead of the maritime threat to an island nation like the United Kingdom did not last long.

World War II

As fighting broke out in 1939, the matchup between the Axis submarine forces and the Royal Navy resembled in many ways the operational situation from the previous World War. While ASW aircraft had advanced a great deal from a performance perspective, they still searched using visual methods and attacked their targets on the surface. German U-boats were remarkably similar to their World War I counterparts in speed, endurance, range, and size. The only significant change was the performance and capabilities of their torpedoes.

While World War II saw airborne ASW operations take place in most theaters, we will restrict our discussion to Allied operations in the North Atlantic. The Allies did not carry out extensive submarine operations in the European theater, and as a result, the Axis air forces and navies did not field large numbers of ASW aircraft. While the United States carried out a highly effective and widespread submarine campaign against Japan, the Imperial Japanese Navy and Army were not able to effectively employ their ASW aircraft. The Battle of the Atlantic offers the best historic case study in wartime airborne ASW.

A detailed discussion of maritime patrol operations during World War II is beyond the scope of this work. It is important, however, to emphasize four key events that occurred during the conflict. First, submarines shifted from operating primarily on the surface at the start of the conflict to operating underwater for long periods of time by the end of the war. Second, World War II witnessed the introduction of electronic and acoustic sensors, allowing aircrews to detect target boats by means other than visual search. These methods allowed submarines to be detected, tracked, and attacked at night, in poor weather, and while submerged.

Third, ASW aircraft gained the ability to attack submarines while on the surface, while submerging, and while moving underwater. Fourth and last, the conflict witnessed the introduction of tactical electronic warfare to ASW operations. The introduction of radar set off a competition between the Allied ASW forces and German U-boat fleet, as advances in radar led to countermeasures, which in turn led to further radar advances to overcome the newly-deployed countermeasures. This pattern of technological advancement, response, and counter-response would be repeated during future ASW campaigns.

Upon the outbreak of hostilities, the Royal Navy immediately instituted a convoy system. RAF Coastal Command, which was responsible for operating the fleet of maritime patrol aircraft, coordinated closely with the

Royal Navy to carry out convoy patrols. Because of the emphasis on strategic bombing, RAF Fighter Command and Bomber Command received the lion's share of funding, leaving Coastal Command with a fleet of medium bombers and obsolescent flying boats. Still, Coastal Command put their forces to good work, flying open-ocean convoy escorts as well as unarmed "Scarecrow" patrols in coastal waters.

German U-boat operations against merchant shipping began slowly, due to the small number of submarines in service and the fact that the majority of the boats were supporting operations in Norway and mining the approaches to British ports. As larger numbers of U-boats went to sea, the Germans focused on exploiting holes in the British convoy system and attacking at times and places that the merchant ships were accompanied by the fewest number of escorts.

The British and the Germans both established dedicated and well-resourced cryptologic efforts and SIGINT units. The Nazis focused on breaking the codes that encrypted British convoy message traffic. The British focused on intercepting and plotting high frequency radio transmissions from deployed U-boats to commanders ashore. They also attempted to break German military traffic encoded with the Enigma cipher machine. These efforts would culminate in successes by the now well-known code-breaking team at Bletchley Park.

The German submarine fleet was placed under tight operational control by commanders ashore. The British, having learned valuable lessons during World War I, established a highly effective operational intelligence cell at the Admiralty. The Submarine Tracking Room used all-source intelligence data to build a holistic picture of how the U-boat force operated. It also generated predictions of the location and activity of individual submarines.

As the number of German submarines rose, the intensity and number of attacks on merchant shipping in the North Atlantic grew. Initial German tactics focused on submerged daytime attacks. While British active sonar was less effective than hoped, its use eventually pushed the Germans to shift to stalking their prey on the surface during the day, positioning their boats ahead of a convoy at dusk, and attacking on the surface at night. As British counter-tactics improved and the number of surface and air escorts grew, the Germans began to attack in the empty parts of the mid-Atlantic, where there was no air cover and the number of escorts was lowest.

The entry of the United States into the war following the attacks on Pearl Harbor opened the Western Atlantic to submarine operations. The Germans immediately dispatched long-range submarines to the Gulf of Mexico, the Caribbean, and the waters off the eastern seaboard of the United

States. The U.S. Navy and Army Air Corps were completely unprepared for ASW operations. There were very few surface escorts, less than twenty ASW aircraft, and no convoy system for merchant ships in coastal waters. In addition, convoluted C2, a complete lack of standard ASW tactics, and non-existent civil defense measures allowed the U-boats to enjoy a very permissive tactical environment. German crews turned the American littoral into a killing field, as hundreds of ships were sunk in the first few months of 1942.

Alarmed at the extremely high number of sinkings off the eastern seaboard, British ASW experts traveled to the United States in the spring of 1942 to meet with their American counterparts. While initially met with skepticism and a measure of hostility, the British officials were able to convince their allies to adopt a coastal convoy system and overhaul their C2.[17] These efforts resulted in the U.S. Navy standing up the ASW Operations Group (ASWORG), which mirrored the highly effective operations analysis research work on ASW being done by the British.[18] The ASWORG recruited extensively from academia and conducted pioneering work in the discipline of operations analysis.[19]

As American ASW efforts improved in the Western Atlantic, German submarines withdrew to the waters east of the Canadian Maritime Provinces and south of Iceland, exploiting the remaining gaps in Allied air coverage. During this period, the Allies remained focused on the strategic bombing campaign in Europe. As a result, the vast majority of large aircraft that rolled off production lines were assigned to strategic bombing units, rather than the maritime patrol squadrons that desperately needed long-range ASW aircraft to provide air coverage at all points along the convoy routes.

This portion of the war witnessed important advances in sensor and weapons technology. The first airborne radar sets flew successfully, allowing ASW aircraft to detect surfaced submarines beyond the visual range of observers, at night, and in poor weather. Coastal Command fielded a powerful searchlight known as the "Leigh Light" that could be slewed to radar contacts and allow night attacks against U-boats. Magnetic anomaly detection (MAD) equipment was introduced aboard ASW aircraft and airships. To help airships discriminate between naturally-occurring magnetic disturbances and actual submarines, engineers developed expendable hydrophones that could float in the ocean and transmit signals to an aircraft overhead. These first sonobuoys were eventually introduced in fixed-wing aircraft as well. Last, American scientists developed and deployed the first airborne LWT, which could be dropped from an aircraft and independently home on a submerged submarine using passive sonar.

The introduction of airborne ASW radars launched a technical race between the Allied and Nazi scientists. German commanders were initially unsure of what was causing an increase in the number of nighttime attacks on U-boats until a British ASW aircraft equipped with radar crashed in France. This clue spurred the Germans to quickly design and deploy a radar warning receiver (RWR). By 1942, these RWR sets were deployed aboard many U-boats, allowing submarine crews to detect an approaching aircraft and submerge before the aircraft could close the range and attack.[20]

Frustrated that airborne radars were becoming less effective and that aircrews were recording fewer sinkings, the Allies deployed new radar equipment that operated at a higher frequency and with a shorter wavelength. This in turn drove the Germans to deploy new RWRs tuned to detect the higher frequency signals.

The Allies also began operating dedicated escort carriers, which were equipped with air wings specialized for ASW patrol flights. In addition to convoy escorts, these carriers were employed very effectively in offensive ASW operations, hunting down and destroying high-value targets such as German tanker submarines.[21] These operations marked the first time a concerted effort was made to rapidly disseminate SIGINT, electronic intelligence (ELINT), and cryptographic information to accomplish a tactical ASW task.[22]

By 1943, sufficient numbers of long-range maritime patrol aircraft existed to allow Allied forces to carry out widespread offensive ASW operations. Planners focused their efforts on the Bay of Biscay, which the majority of German U-boats passed through during their transit from bases in occupied France to patrol areas in the Western Approaches, North Atlantic, and the eastern seaboard of the United States. The Bay offered the highest density of submarine targets other than the waters near the convoys themselves.

The planners intended to maximize detection opportunities, since higher numbers of detections would result in a great number of attacks. A larger number of attacks would mean more submarines sunk or forced to return to port to repair battle damage. Even if the boats were not damaged, higher numbers of attacks would force U-boats to submerge more often, slowing their transit to their patrol boxes. U-boats might also be forced to take longer routes to their operating areas. Taken together, these outcomes would result in the average boat spending more time in transit and less time hunting, which would ultimately mean fewer merchant ships sunk.[23]

Allied operations analysts identified a "ribbon" of water that was the most likely area on average for patrol aircraft to sight a U-boat. The zone was patrolled day and night. Aircrews had modest results at first but eventually sighting and sinking rates steadily increased.[24] The success of the Bay of Biscay offensive eventually forced U-boats to operate primarily underwater, surfacing only when required to recharge their batteries. It also drove the Germans to adopt the snorkel, an apparatus that extended from the conning tower and above the surface to allow the boat to run its diesel engine and ventilate while remaining submerged.

By 1944, the industrial might of the Allies had tipped the scales decisively in their favor. Large numbers of well-equipped and well-trained surface escorts and ASW aircraft savaged the U-boat fleet during the final months of the Battle of the Atlantic. The tactical environment had become so deadly for the German submarine force that more submarines were being sunk than were merchant ships.

In the final months of the war, the Germans attempted to introduce radical new submarine technology. A new, highly advanced submarine design known as the Type XXI went to sea in 1943. The Type XXI featured capabilities that define the modern SSK. It had a streamlined hull optimized for underwater speed and maneuverability rather than prolonged surface operations. It had a high underwater speed and excellent submerged endurance. It had large-capacity batteries, a snorkel, a tactically-effective passive sonar suite, and automated torpedo-handling equipment.

The Germans had also developed radio technology that was poised to thwart the highly effective Allied HF/DF intercept system. The Nazis had perfected a burst transmission system known as "Kurier" that compressed a normal 20-second transmission into a signal that lasted less than a second. Such a short signal could not be intercepted and properly plotted by Allied SIGINT equipment.

Kurier was poised to deny the Allies the highly effective cueing that alerted ASW forces to the presence of U-boats and provided an initial search location. Widespread introduction of the snorkel would drastically degrade airborne radar, which had become the primary search sensor for Allied ASW aircraft. Between these two advances, the Allies faced the prospect of losing their primary methods of ASW cueing and search. Such a loss could cripple Allied ASW forces and deliver the submarine a measure of the invulnerability it enjoyed during World War I.[25]

Luckily for the Allies, Germany was unable to field the Type XXI in sufficient numbers to undermine the effectiveness of ASW operations.[26] As the war ended, the Americans and British were lucky to have concluded the conflict just before their decisive ASW advantage evaporated. As the victors and the vanquished worked to piece back together a burning and broken world, the stage was set for a new geopolitical struggle between the democratic nations of Western Europe and the Americas and the communist Soviet Union and its allies. The United States and its allies enjoyed unchallenged sea control. The Soviet Union sought to guard its territory with powerful armies and to contest the oceans using asymmetric means such as the submarine.

The violent years between 1939 and 1945 witnessed several important advances in ASW and submarine construction that would shape naval operations during the Cold War. First, submarines had shifted from operating primarily on the surface as submersible torpedo boats to moving primarily underwater. Second, engineers had developed a variety of electronic sensors to detect submarines on as well as below the ocean's surface. While radar had proven itself as the primary airborne ASW search sensor, its effectiveness was greatly reduced by the introduction of the snorkel. Sonobuoys and MAD equipment allowed aircraft to detect and track submerged boats. Third, weapons had been developed that allowed an ASW aircraft to attack a submarine on the surface or underwater. Last, World War II had witnessed the use of sophisticated and highly effective operations analysis techniques. These analytical methods and highly effective C2 systems would feature prominently in Cold War efforts to locate and hold at risk the Soviet submarine fleet.

Notes

1. Pre-war exercise summaries and lessons learned attest to the growing threat of submarines and their increasing sea-keeping abilities, range, and lethality. However, senior political and military leaders were uninformed of how these new weapons might pose serious threats to established warfighting techniques. See Jan S. Breemer, *Defeating the U-boat: Inventing Antisubmarine Warfare* (Newport: Naval War College Press, 2010), p. 13.

2. Ibid., pp. 10–12.

3. Alfred Price, *Aircraft Versus Submarine in Two World Wars* (Barnsley, South Yorkshire: Pen & Sword, 2004), pp. 2–5.

4. The German submarine *U-9* torpedoed and sank three British cruisers, HMS *Aboukir*, HMS *Cressy*, and HMS *Hogue*, in less than two hours on 22 September 1914, 1,459 British sailors perishing that day in the North Sea. The incident shocked the Royal Navy, and starkly revealed the lethality of the submarine as an anti-surface warfare weapons system. For more details on this engagement, see Robert K. Massie, *Castles of Steel: Britain, Germany, and the Winning of the Great War at Sea* (New York: Random House, 2004), pp. 133–5.

5. The German airship *L-9* sighted and attacked the British submarine *E-11* on 25 December 1914 in waters off the island of Nordnerney, in North Germany. The *E-11* escaped the encounter unharmed.

6. Breemer, *Defeating the U-boat*, pp. 18–19.

7. Dwight R. Messimer, *Find and Destroy: Antisubmarine Warfare in World War I* (Annapolis, MD: Naval Institute Press, 2001), p. 17.

8. Admiral Reinhard Scheer, IGN (ret.), *Germany's High Sea Fleet in the World War* (London: Cassell, 1920), p. 257.

9. Sir Julian S. Corbett, *Naval Operations* (London: Longmans Green & Co., 1923), vol. 3, p. 140.

10. The British flew periodic ASW patrols in the waters off the main German submarine base at Heligoland Bight. This enabled frequent sightings and the chance to attack or force boats to submerge. Price, *Aircraft Versus Submarine*, pp. 16–18.

11. Bryan Clark and John Stillion, "What it Takes to Win: Succeeding in 21st Century Battle Network Competitions," (Washington, D.C.: Center for Strategic and Budgetary Analysis, 2015), pp. 22–3.

12. Breemer, *Defeating the U-boat*, p. 41.

13. Price, *Aircraft Versus Submarine*, p. 15.

14. Ibid., p. 19.

15. One of the most popular "Scarecrow" aircraft, the British D.H. 6, was inexpensive and simple to fly. It also possessed another trait that made it a natural fit as an early maritime patrol aircraft: it floated well. The unreliable engines of the day meant that powerplant failure was quite common. D.H. 6s were often equipped with inflated bladders to add buoyancy. After being ditched in the ocean, several of these aircraft floated for hours before their pilots were rescued. Ibid., pp. 19–20.

16. Ibid., pp. 34–5.

17. For a summary of the sorry state of U.S. ASW forces upon entry into the war and the success of the U-boat fleet on the eastern seaboard, see Michael Gannon, *Operation Drumbeat: The Dramatic True Story of Germany's First U-Boat Attacks Along the American Coast In World War II* (New York: Harper & Row Publishers, 1990).

18. William F. Story, "A Short History of Operations Analysis in the United States Navy (Monterey, CA: Naval Postgraduate School, December 1968), pp. 27–34.

19. The work of the ASWORG was central in formalizing the field we now know as operations analysis. Concepts of search theory, optimum allocation of search effort, and quantifying probability of detection were pioneered during this time period. Readers interested in operations analysis during the war will be well served by reading the following work. Charles M. Sternhill and Alan M. Thorndike (eds.), "Operations Evaluation Group (OEG) Report Number 51: ASW in World War II," (Washington, D.C.: Chief of Naval Operations, 1946).

20. Clark and Stillion, "What it Takes to Win," pp. 17–19.

21. German tanker submarines were very effective at boosting the number of U-boats at sea. By providing fuel and supplies, they allowed the attack submarines to stay on patrol longer. This resulted in more boats in the convoy lanes at any one time and resulted in higher numbers of sinkings. It also allowed crews to avoid the risky transit through the Bay of Biscay back to port to resupply, lowering the number of U-boats attacked or sunk by Allied ASW aircraft. Allied planners realized the effect of the tankers and were willing to risk revealing the source of their intelligence to hunt down and target these boats at their proscribed rendezvous locations. Brian McCue, "U-Boats in the Bay of Biscay: An Essay in Operations Analysis" (Washington, DC: National Defense University Press, 1990), pp. 32–3.

22. Earlier in the conflict, a great deal of SIGINT and ELINT information was withheld from tactical ASW forces to protect the extremely sensitive sources and methods such as the British ULTRA cryptanalysis program. Towards the end of the conflict the prospect of Allied victory was not in doubt. As such, while Allied signals and electronic

intelligence was very useful, it was no longer a "war winning" capability. As a result, analysts were more willing to use this intelligence information to provide cueing and support tactical employment of ASW ships and aircraft.

23. British and American analysts developed a holistic picture of the manner in which German U-boats crews trained, deployed, operated in a patrol area, transited home, and completed maintenance. This detailed knowledge allowed the identification of highly probably areas to patrol. For details of ASW operations modeling, see McCue, "U-Boats in the Bay of Biscay," pp. 99–108.

24. For details of offensive airborne ASW operations in the Bay of Biscay, see John Campbell, *Royal Air Force Coastal Command: A Short History of the Maritime Air Force Which Protected the United Kingdom's Shipping During World War I and World War II* (Cirencester, Gloucestershire: Mereo, 2013), pp. 155–6.

25. Clark and Stillion, "What it Takes to Win," pp. 19–21.

26. While 139 Type XXIs were preparing to go to sea, only one had made a war patrol by the time the war ended in May 1945. Norman Polmar and Thomas B. Allen, *World War II: The Encyclopedia of the War Years 1941-1945* (New York: Dover, 2012), p. 821.

Chapter 5

Airborne ASW History: The Cold War to the Present

The superpower confrontation that began in the late 1940s placed great focus on submarines and ASW. The United States and its allies sought to counter the military threat of the Soviet Union and its Warsaw Pact partners while bolstering allied and non-aligned countries against the expansion of Communism. While the Warsaw Pact boasted more powerful conventional land forces, the NATO militaries enjoyed a strong measure of naval superiority.

The Soviet Union focused much of its defense spending on building its army and air forces. Rather than seeking a force-on-force naval contest with NATO, the Soviets sought to undermine Western sea control with less expensive, asymmetric systems. Submarines were a logical choice to challenge NATO fleets. In the case of any conventional conflict, NATO would be forced to maintain sea lines of communication between North America and Western Europe. Much like the German U-boat force before it, the Soviet submarine fleet would be well positioned to cut these links and disrupt the flow of troops and material to Europe as well as defend the maritime approaches to the Soviet Union from NATO strike forces intent on attacking land targets.

As the atomic arms race accelerated, the stealth and mobility of submarines made them logical delivery vehicles for nuclear weapons. Given the relatively short range and low cruising speeds of post-World War II bomber aircraft, the submarine seemed to be an excellent launch platform for cruise missiles. Able to approach an enemy coast undetected and close to striking range, the submarine morphed from a tactical weapons platform into a strategic one. This shift in role increased the focus placed on ASW.

The tactical balance between submarines and aircraft found Western airborne ASW forces at a distinct disadvantage during the early years of the Cold War. While new technology such as sonobuoys and MAD had been developed, radar remained the primary search sensor to find submarines. Unfortunately for ASW aviators, the introduction of the snorkel had greatly undermined the effectiveness of airborne radar sets. So too had the development of high frequency burst transmissions. Patrol aircraft needed cueing information to know where to begin to hunt. Without high frequency cueing, airborne ASW search effectiveness was greatly decreased. These factors left NATO searching for new approaches to submarine detection.

As a result, NATO turned to acoustic sensors to regain a reliable search capability. Running submerged on their electric motors, submarines were very quiet and difficult to detect with passive sonar hydrophones. However, boats still needed to rise to periscope depth to extend their snorkel, start up their diesel engines, and recharge their batteries. Submarine diesel engines were mounted directly to the hull, so their vibrations transferred to the hull and into the ocean.

This large amount of sound was readily detectable. The acoustic energy also tended to cluster at distinct and predictable frequencies, based on the technical characteristics of the submarine and the diesel engine. These frequencies of narrowband noise were called "tonals" by engineers and aviators. Using signal processing techniques, narrowband tonals could be isolated from the broadband background noise of the ocean. This allowed aviators to drop passive narrowband sonobuoys and pick up the telltale sound of a snorkeling submarine at long ranges.

In order to refine the location of the target enough to establish a precise track or allow a torpedo attack, the aviators had to shift from using long-range narrowband signal processing to dropping short-range sonobuoys to pick up the more diffuse broadband noise produced by the submarine. Early sonobuoy systems and acoustic processors were difficult to operate. Oftentimes, a submarine would stop snorkeling before an aircrew could complete the lengthy and complicated task of converting long-range narrowband contact to short-range broadband contact.[1] As a result, Western ASW crews in the early 1950s were often successful at detecting a snorkeling boat, but struggled to localize and track the submarine once it shifted back to its quiet electric motors.

In 1954, the United States launched USS *Nautilus* (SSN-571), the first nuclear-powered submarine. Nuclear power freed *Nautilus* from the requirement to operate near the surface. The boat could dive deep and operate at high speed for as long as her crew desired. *Nautilus* was fast and maneuverable, so much so that the active sonars of the time could not remain trained on her position long enough to track her.[2]

However, *Nautilus* and other early nuclear submarines had a critical flaw, one that would tilt the ASW advantage from the submarine back to the airplane. In order to circulate coolant and remove massive amounts of heat from the reactor, nuclear boats required large coolant pumps. These boats also used large steam turbines, generators, and reduction gears, all of which were mounted on the hull much like the diesel engines of conventional submarines were. As a result, these early nuclear boats generated very loud and predictable tonals. A conventional submarine could shut its diesel engine down and operate on batteries, becoming quiet in the process. Nuclear submarines, on the other hand, were obliged to run their coolant pumps at all times. The combination of high radiated noise levels and precise tonals made early nuclear submarines ideal targets for the well-developed narrowband acoustic processing techniques practiced by NATO aviators.

The speed, range, and submergence capabilities of nuclear submarines had important tactical implications. The LLOA concept forced conventional submarines to get out in front of a target before launching a torpedo attack. As a result, ASW aviators had traditionally focused on patrolling the sectors ahead of and off the beams of a HVU or convoy. Because SSNs were as fast as the majority of their targets, ASW aircrews had to shift their focus and stay alert for boats approaching the rear of a convoy.

In addition to narrowband sonobuoys and acoustic processing techniques, the United States developed and deployed another important sensor system during the 1950s. Experiments before and during World War II had revealed the existence of a sound ducting phenomenon known as the Deep Sound Channel (DSC). The DSC allowed sound to travel across hundreds or even thousands of miles and still be picked up by hydrophones mounted deep in the ocean.

The U.S. Navy deployed a series of deep ocean hydrophone arrays known as the Sound Surveillance System (SOSUS). Using narrowband signal processing and analysis techniques, technicians were able to detect and plot the location of ships and submarines, including Soviet boats. When a signal was picked up and classified as a foreign submarine, SOSUS technicians would identify a Submarine Probability Area (SPA), which was forwarded to tactical ASW units. These task forces would then assign ships, submarines, and patrol aircraft to search the SPA and attempt to detect the submarine operating nearby.

SOSUS gave the United States and its allies the ability to accurately and effectively plot the location of nuclear submarines. In a similar manner, narrowband passive sonar gave surface ships, submarines, and aircraft the ability to detect and track Soviet boats at long ranges. By the late 1950s, an effective concept of operations had been devised to deal with the threat of the Soviet submarine fleet. SOSUS would serve as an indications and warning, intelligence gathering, and cueing system. Land-based ASW aircraft would be dispatched to sanitize SPAs and develop tracks of Soviet boats. These boats would either be tracked and held at risk by the patrol aircraft or handed off to friendly SSNs to be tailed for extended periods.[3]

The U.S. Navy established forward deployment sites for patrol aircraft and a series of ASW operation centers worldwide. This infrastructure was augmented by air bases, patrol aircraft squadrons, and C2 facilities in NATO countries. In the case of war, task forces and merchant convoys would have air cover provided by land-based patrol aircraft and ASW aircraft based aboard escort carriers. Eventually, the high cost and specialized nature of escort carriers led to their retirement. The specialized ASW squadrons embarked aboard the escort carriers were transferred into the multi-mission air wings of larger attack carriers.

By the early 1960s, submarine-launched ballistic missiles (SLBMs) had been developed and deployed by both NATO and the Soviet Union. The limited range of these missiles forced both sides to deploy their SSBNs near the adversary's territory. The Soviet Union deployed SSBNs off the coast of the United States in patrol boxes in the North Atlantic and North Pacific. NATO developed the ability to routinely detect, localize, and track these SSBNs during their forward patrols. This was often accomplished using cueing data provided by SOSUS, initial search and detection made by an ASW aircraft, and extended tailing by an SSN. American patrol aircraft were often used to augment fast attack boats when a trail went cold or to provide redundancy when the SSBN moved into the "alert" phase of its patrol.[4]

The Era of Western Acoustic Advantage

The United States was quick to recognize the vulnerability of its SSNs and SSBNs to narrowband acoustic detection. During an early patrol of an American SSBN, SOSUS arrays picked up the boat's tonals from several thousand miles away, tracking the submarine as she sailed from the eastern seaboard of the United States to the United Kingdom.[5] As a result, the U.S. Navy began an aggressive submarine noise-reduction program. New submarine designs featured a series of flexible mounts that isolated the powerplant from the hull itself. These mounts were called "rafts," and they served to drastically reduce the magnitude of the vibrations that were transferred from the rotating pumps and machinery to the hull and into the water.

Rafting resulted in sharp decreases in the radiated noise levels of American submarines. This made them far less susceptible to detection by long-range sensors like SOSUS and short-range passive sonars. Other NATO navies copied rafting in their own submarines and warships. The Soviets, however, were unaware of how vulnerable their boats were to detection by Western ASW forces. As a result, they were much slower to invest in quieting technology.[6]

The wide disparity in noise levels between Western and Soviet submarines drove NATO to invest heavily in high-performance passive sonars, powerful computers for acoustic processing, and high levels of quality control to keep their submarines quiet. Western ASW aircraft received periodic upgrades to their tactical systems to allow them to deploy more sonobuoys and to process larger amounts of acoustic information.

The high noise levels of Soviet boats allowed NATO ASW aircraft to frequently exploit a long-range sound propagation phenomenon known as the convergence zone. In deep water, variations in temperature and pressure drive sound to bend downwards and then be focused upwards in a series concentric circles spaced about every 30 nautical miles from the target. This allowed aviators to pick up tonals at ranges far greater than might be possible otherwise. Using a methodical localization process, the aviators could eventually determine where the target might be and move to short-range tracking and intelligence gathering.

Convergence zone tactics were important because they allowed ASW aircraft, especially land-based patrol aircraft, to cover very large search areas.[7] This made maritime patrol aircraft highly effective wide-area search assets, provided they were supported with good cueing information such as a SPA or an ELINT intercept.

During this time period, the U.S. Navy deployed another critical tool for ASW search. Since the end of World War II, operations analysts and applied mathematicians had been studying the problem of finding lost objects in the ocean and how to refine a search to take new information into account. After success in two high profile search and salvage operations during the late 1950s, the Americans realized that applied mathematics could be used to predict the location of Soviet submarines and generate search plans that would yield the best chance of success given a set number of ships, submarines, and aircraft.[8]

In the early 1970s, the U.S. Navy began using a series of computer assisted search (CAS) planning tools. These tools used mathematical functions to describe the likely motion parameters of a Soviet submarine such as course, speed, and depth. Given a piece of cueing information, the CAS system would generate a probability map of where the target was most likely located and provide a recommended search plan.[9]

CAS systems needed accurate data on how target submarines tended to operate. Luckily for NATO, its aircraft and SSNs had been engaged in covert trails of Soviet boats for over a decade. Analysts had reams of contact reports listing position, course, and speed of target submarines. This rich data set paired perfectly with CAS techniques and the rapidly improving computers of the time.

As the aviators flew their missions, positive, negative, and ambiguous contact data would flow back into an operations center. The data would be assigned a confidence value and then entered into a computer database. The CAS system would update the probability function and modify the search in near real time.[10] These systems were highly effective, leading to large improvements in detection rates during exercises and real-world prosecutions.[11]

By the mid-1970s, NATO could proudly claim a highly effective airborne ASW force. SOSUS offered a reliable and actionable cueing system. CAS systems provided optimum search plans. Patrol aircraft were able to arrive quickly at a SPA, employ a large sonobuoy search pattern, and detect their target, usually using convergence zone tactics. This enabled the Western navies to hold target boats at risk or gather intelligence.

Geographic factors and operational considerations drove various NATO air arms to adopt different ASW roles. Norway sat next to the training areas and transit lanes that the Soviet Navy used to access the North Atlantic. Their ASW aircraft were often first to arrive over a Soviet boat heading outbound for a patrol. The British RAF and Royal Navy were responsible for defending the Western Approaches during any conflict in Europe. They operated frequently against boats transiting through the Greenland, Iceland, and United Kingdom (GIUK) Gap. These waters were frequently patrolled by Soviet diesel submarines, which allowed the British to develop a strong competency against conventional submarines in shallow water.[12]

The Canadian Armed Forces frequently operated against forward-deployed Soviet SSBNs.[13] They were joined by their American compatriots, who kept tabs on Soviet missile submarines as they traveled to and from their patrol boxes off Newfoundland and Bermuda in the Atlantic and their operating areas in the Northern Pacific.[14]

The Soviet Union developed its own airborne ASW force as well. Western submarines had grown very quiet and a large part of the NATO submarine force was nuclear-powered, which reduced the effectiveness of airborne radar. While the Soviets flew with sonobuoys and periscope-detection radars similar in concept to Western equipment, they were forced to investigate other methods of submarine detection. The Soviets were able to call on their numerous design bureaus and well-funded applied science and research establishment to investigate non-acoustic submarine detection methods.[15]

Very little information on Cold War Soviet maritime patrol operations is available in the open source literature. However, it is clear that the Soviets put great effort into trying to detect the internal waves and wakes generated by Western submarines as they moved through the deep ocean. The Soviets investigated and tested light detection and ranging (LIDAR) systems aboard their maritime patrol aircraft, which could in theory detect submarines operating close to the surface.[16] The Soviets carried MAD gear and may have attempted to pick up the changes in magnetic fields caused by disturbed ions in submarine wakes.[17] Unfortunately, there is very little information available on how the Soviets approached airborne ASW C2. Information on tactics, sensor effectiveness, and overall capability of Soviet aircraft to detect and track NATO submarines is not readily available in open sources.

The 1970s and early 1980s were happy and productive times for Western ASW forces and ASW aircrews in particular. However, by the mid-1970s, the Soviets had become aware of just how vulnerable their submarine fleet was to detection and tracking by NATO ASW forces.[18] A spy ring inside the U.S. Navy had passed the Soviets codebooks, message traffic, and information on the capabilities of NATO ASW forces for several years. The Soviets embarked on a radical quieting and submarine redesign effort to wrest away the enormous advantages in stealth and sensor performance that NATO enjoyed.

The Era of Acoustic Parity

In 1979, the Soviets deployed their first submarine that carried an effective sound-isolation system. While the vast majority of the Soviet fleet didn't utilize rafting and remained very noisy, this was a sign to NATO that their acoustic advantage was on the brink of being eroded. Through the early 1980s, the Soviets deployed increasing numbers of SSNs, SSGNs, and SSBNs that incorporated effective quieting technology while decommissioning older, noisier boats.

Rafting, improved quality control, and advances in propeller design lowered the noise levels that Soviet boats put into the water. These quieter signals in turn lowered the range at which SOSUS could detect target submarines. Fewer, shorter-range detections made cueing information less reliable and less available to ASW aircrews. Lower radiated noise levels also hurt aircrews' ability to search. The quieter the boat, the lower the chance that NATO aviators had of employing long-range convergence zone tactics. This shrunk the area that patrol aircraft could effectively search while still having a realistic change of detecting their target.

Soviet missile design had improved as well. This allowed increases in SLBM range, which in turn meant that Soviet SSBNs didn't need to come as close to American shores to hold their targets at risk. Soviet missile submarines could hide under the ice shelf in the Arctic or even stand alert in port rather than run the gauntlet of NATO ASW forces in the Norwegian Sea, GIUK Gap, and operating areas of the North Atlantic.

The Soviets also designed several classes of submarines that incorporated novel design features. These boats utilized titanium hulls to allow them to operate much deeper than Western submarines. They boasted novel powerplants that used liquid metal instead of pressurized water as a reactor coolant and they were constructed with tough double hulls that gave a large amount of reserve buoyancy to enhance survivability against torpedoes.

By 1984, the Soviets had introduced the first two classes of SSNs that could routinely evade SOSUS.[19] This capability greatly alarmed NATO. If SOSUS ceased to be effective, tactical ASW forces would lose the critical cueing information they needed to start a search. Quieting levels had decreased the detection ranges of tactical sonobuoys to the point that effective wide-area airborne search with ASW aircraft might no longer be realistic.

The U.S. Navy responded with several initiatives to meet the new Soviet submarine threats. First, they began deploying a class of auxiliary oceanographic ships equipped with a hydrophone array known as the Surveillance Towed Array Sensor System (SURTASS). These vessels could be deployed to known or suspected Soviet operating areas and act as mobile SOSUS arrays. This would restore some of the effectiveness of the NATO cueing system.

The United States also began investigating sonar arrays that would use a large number of short-range sensors instead of the small number of very large, long-range SOSUS arrays. The Fixed Deployable System (FDS) would feature dozens of passive hydrophones deployed in choke points. Instead of using the DSC, FDS would use a phenomenon known as Reliable Acoustic Path (RAP), where temperature and pressure gradients bend sound downwards from a shallow submarine to a deep sensor lying on or near the seabed.[20]

The U.S. Navy also began work on two new sonobuoy systems. The first, known as the Tactical Surveillance Sonobuoy (TSS), featured large batteries and a much longer life than a normal passive sensor. Rather than a typical eight-hour passive sonobuoy life, TSS buoys could operate for five to seven days at a time.[21] An ASW aircraft would seed a suspected operating area with TSS sensors and wait. While each buoy had a short detection range, it was very unlikely that a Soviet submarine patrolling in the area would not wander into detection range of one of the many buoys.

The Americans also turned to using powerful active sonar pulses to detect quiet Soviet boats. While new-build submarines were very quiet, they remained quite large, and were readily detectable with active sonar. During the 1950s, the U.S. Navy had utilized small explosive sound sources to generate active sonar emissions.[22] Now, 30 years later, the Americans returned to the technology, hoping that advances in computing power and training could make the so-called Explosive Echo Ranging (EER) system a success.

The new generation of fast, deep-diving, double-hulled Soviet boats had rendered many of the LWT designs of the era obsolete. The U.S. Navy fielded a new powerful air-launched torpedo that was fast enough and could dive deep enough to catch the new Soviet boats. The weapon boasted a shaped-charge warhead that could cut through double hulls and utilized improved digital signal processing to help filter out interference of submarine-launched decoys.[23] The British, French, and Italian navies also developed their own new LWT designs with similar capabilities.[24]

Last, NATO worked to improve C2 procedures and operations analysis capabilities. By 1984, a CAS system was in place in every American ASW operations center worldwide.[25] The newest of the U.S. Navy systems could build search plans for five individual submarines while generating employment plans for friendly ships, submarines, aircraft, and SURTASS vessels.[26] Research was undertaken to integrate CAS systems and tactical decision aids onboard ASW aircraft themselves, assisted by the smaller size and higher processing power of new computers coming online.[27]

Much like the situation faced by the Allies in the final months of World War II, by the late 1980s NATO stood at the brink of losing its hard-won ASW superiority. However, in the winter of 1991, the Soviet Union suddenly ceased to exist. The massive Russian submarine fleet was confined to port as the Russian Federation lacked the money to maintain its submarines and pay its sailors. Almost in an instant, the threat of the Soviet fleet vanished and along with it, the impetus to maintain a well-funded and highly trained ASW force. The ensuing years would see NATO and other Western nations squander their hard-won ASW experience, force structure, and warrior culture.

The Period of Decline

The end of the Cold War led to very large cuts in the defense budgets of NATO nations. The United States in particular reoriented the focus of its armed forces from fighting nuclear and conventional conflicts against a superpower to responding to security contingencies and unrest in unstable areas of the world.

This shift from preparing for superpower conflict to meeting regional unrest had great effect on "high end" conventional warfighting platforms. Maritime patrol aircraft and the C2 organizations that directed Theater ASW (TASW) operations were some of units that were hit hardest by spending cuts. During the 1990s, the U.S. Navy slashed its fleet of active maritime patrol aircraft from twenty-four squadrons to twelve. Operations centers the world over saw their CAS systems de-funded and shut down and their well-trained personnel retired or assigned to different roles.

Air arms worldwide shifted their maritime patrol aircraft from traditional ASW operations to missions more relevant to the changing security environment. These new assignments included counter-narcotics, overland Intelligence, Surveillance, and Reconnaissance (ISR), and patrol operations intended to track merchant shipping and to combat smuggling. These new missions diluted the focus of maritime patrol crews, forcing them to master several dissimilar missions instead of training primarily for ASW operations. Aviators had far fewer chances to fly "on top" of actual submarines and learn how to conduct real-world ASW. This led to sharp reductions in critical skillsets, familiarity with tactical oceanography, and overall airborne ASW readiness. It also eviscerated the professional culture that emphasized robust operations analysis, highly effective planning ashore, and tight C2 relationships.

Despite the collapse of the Soviet Union and the fact that the Russian submarine fleet was largely confined to port, the submarine threat worldwide did not disappear, but rather morphed into a new form. During the later years of the Cold War, European navies built and deployed a variety of very quiet, capable, and relatively inexpensive SSKs.

Littoral operations proved a steep challenge for ASW systems designed and fielded during the Cold War. Western ASW forces had been tasked primarily to go after Soviet boats operating in the open ocean. These operating areas featured relatively uniform water columns, deep water, and low levels of merchant shipping. The littoral areas where SSKs were frequently employed were close to shore, noisy, and frequently shallow. From a tactical oceanographic perspective, these areas were highly variable and very harsh environments in which to conduct ASW operations.

Passive sonar was often marginally effective in these littoral areas, especially against quiet, modern SSKs. Multistatic sensors like EER had been optimized for homogenous parts of the open ocean and were not particularly effective in shallow water either. Weapons developed to attack deep-diving

Soviet submarines were poorly suited to the harsh acoustic environment in shallow water. In short, the ASW sensor and weapons portfolio that had been highly effective during the Cold War was poorly suited to the current submarine threat.

As economic growth and the forces of globalization lifted formerly poor nations to new heights in the 1990s and 2000s, politicians sought to put their nations' newfound wealth to work and improve their military capabilities. Many nations bought advanced SSKs. Dozens of inexpensive and capable submarine types such as the Russian Project 877/KILO class, French "Scorpène" class, German Type 209 and Type 214 classes, and Swedish *Västergötland* class proliferated worldwide.

Western airborne ASW proficiency and combat readiness reached its nadir in the late 2000s. Years of war in the Middle East and Southwest Asia had drawn resources away from traditional maritime missions. American and European patrol aircrews focused on flying overland ISR operations. In 2007, more than a quarter of the U.S. Navy's fleet of patrol aircraft was grounded due to structural fatigue.[28] This left barely enough aircraft for operational missions overseas, to say nothing of training sorties or ASW exercise flights. ASW proficiency plummeted, with junior operators having little opportunity to master their craft and senior personnel little time to impart their knowledge on their replacements.

In 2010, the Nimrod MR4A, replacement for the RAF's fleet of MR2 aircraft, was canceled due to delays, cost overruns, and cuts to defense spending. The year later, the remainder of the Nimrod fleet was retired, leaving the United Kingdom without any maritime patrol aircraft. The Nimrod community was viewed by many ASW aviators as boasting some of the most skilled and experienced operators in the world, due to their low rate of personnel turnover, high aircrew stability, and extensive operational experience.

For two decades after the collapse of the Soviet Union, Western airborne ASW forces suffered greatly. Funding cuts eviscerated force structure. A lack of threatening adversaries removed the necessity to perform effectively on station and robbed aviators of valuable operational experience. A generation of aviators grew up and retired without honing their craft and passing down critical knowledge. What little funds were available for airborne ASW research and development focused on improving torpedo and multistatic sonobuoy performance in shallow water.

Despite significant advancements in computing power, operations analysis systems did not transition from operations centers ashore to tactical systems onboard the ASW aircraft themselves. Knowledge of operations analysis and tactical oceanography suffered greatly during these years. It would take the rise of a mature and capable Chinese submarine fleet and a resurgent Russian submarine force to refocus Western navies on the importance of airborne ASW.

Resurgence

As the second decade of the twenty-first century began, attention turned to the Pacific Rim. China was growing at breathtaking speed, both economically and militarily. The Japanese defense forces were acutely aware of the military threat posed by their neighbors. The Japan Maritime Self-Defense Force (JMSDF) had long possessed a fleet of very capable and experienced ASW crews. These aviators were among the first to notice the increasing numbers and capabilities of the Chinese submarine force.[29]

An increasingly assertive Russia began deploying submarines at much higher rates. Boats carried out more frequent operations in the Norwegian Sea and North Atlantic. In 2012, Russian boats began operating off the eastern seaboard of the United States, patrolling in waters near American SSBN bases.[30] In 2017, the Russian Navy deployed several boats to the Mediterranean, playing cat and mouse games with European and American warships.[31] Russian Project 636/KILO class submarines operating from a forward naval base in Syria launched repeated land-attack cruise missiles strikes into that country while sailing in the eastern Mediterranean Sea.[32]

After two decades of neglect, navies and air forces began arming themselves with new aircraft, sensors, and weapons to meet the growing submarine threat. The United States, United Kingdom, India, and Australia all began flying the P-8A. Japan introduced the P-1 patrol aircraft to replace its fleet of P-3Cs. China fielded the Y-8Q, marking the first time that nation operated a fixed-wing ASW aircraft.

Sensor technology marched forward as well. The U.S. Navy focused on improving the performance of its multistatic systems by shifting from explosive sound sources to more sophisticated electronic transmitters. Older multistatic systems such as EER and Improved EER featured high false alarm rates in shallow water.[33] Rather than using explosive charges to produce sound impulses, the new American Multistatic Active Coherent (MAC) system used electronic arrays to generate waveforms tailored to the tactical situation and environmental conditions at hand. More advanced sensors and signal processing offered hope that ASW aircraft might recover some measure of advantage over target submarines.

New technologies offered promises as well. Widespread deployment of unmanned aerial vehicles (UAVs) in the early 2000s led navies worldwide to focus on the development of unmanned underwater vehicles (UUVs) and unmanned surface vehicles (USVs). In 2016, the U.S. Defense Advanced Research Projects Agency began development of an USV to track adversary submarines. The ASW Continuous Unmanned Trail Vehicle was intended to track SSKs for extended periods with minimal human supervision. Substituting a relatively inexpensive USV for a SSN or squadron of ASW aircraft seemed an attractive potential solution to meeting the challenge of numerous and inexpensive SSKs.

The U.S. Navy also returned to using long-dwell passive sensors. Research and development work began in 2014 on the Next Generation Airborne Passive Sensor, a sonobuoy launched by ASW aircraft and intended to drift at depths of 15,000 to 20,000 feet for approximately a month.[34] Long-dwell systems would allow ASW aircraft to seed a suspected operating area and wait for a submarine to approach one of the sensors.

Long-dwell systems seemed to offer cost-effective approaches to wide-area ASW search against quiet targets. They also opened up the possibility of updating search algorithms with negative contact information to improve analysis of likely target submarine locations. Open source data suggested that the U.S. Navy recognized that it lacked sufficient doctrine and operations analysis tools to meet the modern submarine threat. The U.S. Navy publicly published the FSASW doctrine that focused on holistically targeting and degrading the submarine kill chain in its entirety.[35] The service also undertook work to recapitalize its tactical decision aids and C2 systems ashore and afloat.

The Future

It is impossible to predict with a high level of certainty what the airborne ASW world of the next few decades will look like. However, surveying the past and examining current technological trends can help us make some rough predictions.

Submarines will continue to grow quieter. Russian designs have already reached acoustic parity with Western ones. While China's nuclear submarine fleet is still relatively noisy, it is likely that new-build Chinese SSNs will feature improved quieting technology.[36] This will force ASW aircraft to employ multistatic sonobuoys, long-dwell passive sonobuoys, and deep-drifting sonobuoys utilizing RAP to achieve required levels of area coverage.

Conventional submarines will likely see improvements in their underwater mobility and endurance. AIP technology has already proliferated to many nations. Advances in battery technology in general and lithium ion batteries in particular will allow SSKs to remain submerged longer and to transit at higher speeds. More time between snorkeling periods will degrade the effectiveness of airborne radar, the traditional search sensor used to detect conventional submarines.

Great advances are being made in the field of autonomy, meaning that UUVs and USVs to are likely to contribute to warfare problems over the coming decade just as UAVs have played a key role in ISR operations during the last two decades. Sensor-equipped UUVs will allow ASW forces to sanitize large search areas and relay environmental information to commanders ashore. USVs may tail submarines on patrol, freeing ASW aircraft and SSNs to focus their efforts elsewhere.

A force of submarines, ships, aircraft, and unmanned vehicles are likely to be networked together to share tactical and environmental information. This will allow forces to fuse information together as well as rapidly build a common operating picture and increase situational awareness. Command centers ashore will likely utilize sensor fusion and "big data" analytics approaches to sift through the ocean, locate signals of interest, and build prosecution plans.

High level of acoustic quieting and improved underwater endurance are poised to make the traditional ASW search sensors of the past 50 years, passive sonar and airborne radar, far less effective. With history as a guide, we can be reasonably confident to say that the next two decades will probably witness air ASW forces working to perfect multistatic active sonar as well as pursue non-acoustic sensor methods such as internal wave detection, long-range MAD equipment, multispectral imaging, and LIDAR. These approaches will probably be combined with human-machine teaming decision support systems, where human decision makers are paired with computers that provide numerous optimized courses of action. The tactician or sensor operator can then pick the option that is best suited to the scenario at hand.

Airborne ASW has always been a contest between the hunted submarine and the hunting aircraft. While we cannot be sure of what the future holds, we can be confident that both sides will compete aggressively to seize a tactical advantage or to claw back from a position of comparative weakness.

Notes

1. Owen J. Cote, Jr. *The Third Battle: Innovation in the U.S. Navy's Silent Cold War Struggle with Soviet Submarines* (Newport: Naval War College, 2003), p. 39.
2. Ibid., p. 21.
3. Bryan Clark and John Stillion, "What it Takes to Win: Succeeding in 21ˢᵗ Century Battle Network Competitions" (Washington, D.C.: Center for Strategic and Budgetary Analysis, 22 January 2015), p. 36.
4. Larry Robideau, "Third Battle of the Atlantic 1962-1991," *Cold War Times* (February 2006), pp. 14–20.
5. An authoritative article published by a former SOSUS engineer describes the very long-range detection of USS *George Washington* (SSBN-598) during a voyage in 1961. Edward C. Whitman, "SOSUS: The Secret Weapon of Undersea Surveillance," *Undersea Warfare Journal*, Vol. 7, No. 5 (Winter 2005).
6. The Soviets were unaware of how vulnerable their boats were to detection and tracking until they analyzed information passed to them by a former Navy communications specialist, John Walker Jr. The realization of their boats' vulnerability can be inferred from the drastic shift from constructing deep diving, high-speed submarines such as the Project 685/MIKE class SSN and Project 661/PAPA class SSGN to the larger, quieter and multi-role SSNs such as the Project 671RTM/VICTOR III, Project 945/SIERRA, and Project 971/AKULA.
7. For a discussion of convergence zone tactics and the ability of American ASW aircraft to sanitize large search areas during operations against Soviet submarines, see Lcdr. Ryan Lilley, USN, "Recapture Wide-Area Antisubmarine Warfare," U.S. Naval Institute *Proceedings*, Vol. 140/6/1,336 (June 2014).
8. Henry R. Richardson, Lawrence D. Stone, W. Reynolds Monach, and Joseph H. Discenza, "Early Maritime Applications of Particle Filtering," *Proceedings of SPIE*, Vol. 5204 (Bellingham, WA: SPIE, 2003), pp. 172–3.
9. Daniel H. Wagner, *Naval Tactical Decision Aids* (Monterey, CA: Naval Postgraduate School, 1989), pp. II-57–II-62.
10. Ibid., p. II-5.
11. J.R. Frost and L.D. Stone, "Review of Search Theory: Advances and Applications to Search and Rescue Decision Support" (Washington, DC: U.S. Coast Guard, 2001), pp. 3–4.
12. For details of RAF Coastal Command Nimrod MR2 aircrews operating against Soviet diesel submarines in the GIUK Gap, see Tony Blackman, *Nimrod: Rise and Fall* (London: Grub Street Publishing, 2013), pp. 74–7.
13. Michael Whitby, "Doin' the Biz: Canadian Submarine Patrol Operations Against Soviet SSBNs, 1983-87" in Bernd Horn (Ed.), *Fortune Favours the Brave: Tales of Courage and Tenacity in Canadian Military History* (Toronto, Ontario: Dundurn Press, 2009), pp. 287–332.
14. Robideau, "Third Battle of the Atlantic 1962-1991," pp. 14–17.
15. Norman Polmar and Edward Whitman, *Hunters and Killers Vol. 2: Anti-Submarine Warfare from 1943* (Annapolis, MD: U.S. Naval Institute Press, 2016), pp. 147–50.
16. The Tu-142 maritime patrol aircraft is reported to have flown with a prototype LIDAR system installed. It appears the system was immature and not operationally effective, due in part to poor signal processing and problems stabilizing the laser. For details on Soviet LIDAR research, see John Olav Birkeland, "The Potential for LIDAR as an Antisubmarine Warfare Sensor" (Glasgow, Scotland: University of Glasgow, 2009), pp. 46–7.

17. "Hunting Submarines with Magnets," *The Economist* (12 November 2016).

18. Clark and Stillion, "What it Takes to Win," p. 40.

19. Ibid., pp. 40–2.

20. Cote, *The Third Battle*, p. 78.

21. "GAO/NSAID-91-41: Tactical Surveillance Sonobuoy" (Washington DC: U.S. General Accountability Office, January 1991), p. 2.

22. Roger A. Holler, "The Evolution of the Sonobuoy from World War II to the Cold War," *U.S. Navy Journal of Underwater Acoustics* (January 2014), pp. 332–4, http://www.dtic. mil/dtic/tr/fulltext/u2/a597432.pdf (Accessed January 28, 2015).

23. For details of Soviet submarine developments and the inadequacy of legacy LWTs such as the American Mk 46, see Cote, *The Third Battle*, pp. 59–60.

24. The British Sting Ray and French/Italian MU90 Impact are good examples of modern high-speed LWTs equipped with shaped-charge warheads.

25. Wagner, *Naval Tactical Decision Aids*, p. II-61.

26. Ibid., p. II-36.

27. Ibid., p. II-21.

28. Jeff Schogol, "P-3C Orions Grounded," *Stars and Stripes* (18 December 2007).

29. Japan Protests to China Over Incursion by Nuclear Sub,' *Washington Post* (13 November 2004).

30. Bill Gertz, "Russian Sub Skirts Coast," *Washington Free Beacon* (5 November 2012); David E. Sanger and Eric Schmitt, "Russian Ships Near Data Cables are too Close for US Comfort," *New York Times* (25 October 2015).

31. Dave Majumdar, "The U.S. Navy is Trying to Track Down 'Carrier-Killer' Russian Nuclear Submarines in the Mediterranean," *The National Interest* (10 December 2016), http://nationalinterest.org/blog/the-buzz/the-us-navy-trying-track-down-'carrier-killer'-russian-18704 (accessed 5 July 2018); Julian E. Barnes, "A Russian Ghost Submarine, Its U.S. Pursuers, and a Deadly New Cold War," *Wall Street Journal* (20 October 2017).

32. "Russian Submarines Fire Cruise Missiles at Jihadi Targets in Syria," *Reuters* (22 September 2017).

33. "Coherent Multistatic Active Sonar LRT," U.S. Small Business Administration (2005), https://www.sbir.gov/sbirsearch/detail/307409.

34. "N142-117: Components for a Deep Drifting Sonobuoy," U.S. Navy Office of Naval Research, May 23, 2014, http://www.navysbir.com/n14_2/N142-117.htm.

35. Capt. William J. Toti, USN (ret.), "The Hunt for Full-Spectrum ASW," U.S. Naval Institute *Proceedings*, Vol. 140/6/1,336 (June 2014).

36. "The PLA Navy: New Capabilities and Missions for the 21st Century" (Suitland, MD: Office of Naval Intelligence, 9 April 2015), p. 16.

Chapter 6

Airborne ASW Weapons

To understand the ultimate portion of the airborne ASW kill chain, aviators should be armed with the knowledge of how their weapons function. They should understand the capabilities and limitations of ASW weapons and comprehend how past technical advances drove weapons development. During the last century, airborne ASW weapons have changed a great deal. The competition between submarine and aircraft has often witnessed a new weapon being fielded and a corresponding countermeasure being developed. This competition between weapon and countermeasure has driven weapons technology and employment methodology.

We will not restrict our discussion of airborne ASW weapons to modern equipment alone. Understanding the various types of weapons that were used during different eras of ASW provides context and allows aviators to consider how future technical advances are likely to drive weapons development. Additionally, we will examine the various weapons systems that submarines have used to attack aircraft and finish the chapter with a discussion of submarine countermeasure systems.

Machine Guns and Cannon

During the first half century of ASW flying, aircraft frequently employed machine guns and cannon. Machine guns have been mounted as forward-firing weapons to attack submarines and as turreted defensive weapons to protect ASW aircraft against attacking fighters. Cannon, on the other hand, were employed exclusively in forward-firing configurations to attack surfaced submarines.

The earliest military aircraft in World War I were employed as unarmed scouts, using their mobility and ability to climb to altitude to serve as reconnaissance platforms. However, it was not long before pilots started toting firearms in their kit bags to take pot shots at enemy aviators. During the conflict, technology advanced rapidly, and aircraft were soon flying with forward-firing machine guns and aft-facing, crew-served weapons.

Early ASW aircraft were extremely limited in their ability to carry heavy payloads, and this resulted in early guns being small in caliber. These machine guns lacked the power to cause structural damage to submarines. However, they were useful as anti-personnel weapons to attack submarine crews on deck or in the conning tower as the patrol aircraft approached to drop bombs.

To defend against attacking aircraft, submarines soon took to arming themselves with AAA. As submarines developed an air defense capability, forward-firing guns became crucial defenses for ASW aircraft. Well-placed rounds from a machine gun could suppress fire from submarine gun crews and protect the attacking aircraft as it closed in to deliver its weapons.

During World War II, many medium and heavy bombers were pressed into service as ASW aircraft. Many of these aircraft boasted considerable firepower in the form of forward-firing machine guns and defensive turrets. This heavy armament combined with limited armor protection meant that many World War II ASW aircraft were quite resilient, and able to protect themselves from marauding fighters.

World War II also witnessed light and medium bombers attacking submarines in port or while transiting on the surface close to port. Many of these bombers were equipped with medium or heavy cannon. Unlike the lighter machine guns aboard patrol aircraft, these were powerful enough to penetrate submarine hulls and cause structural damage, not simply act as anti-personnel weapons. Such damage usually did not sink the target boat, but would frequently prevent the submarine from submerging, forcing the boat to abandon its mission.

After the submerged revolution, submarines ceased to spend large amounts of time on the surface. This removed any necessity for submarines to carry AAA or for ASW aircraft to carry machine guns for suppressing fire while they approached to attack. By the mid-1950s, fighter aircraft had become much faster than their World War II predecessors and had begun to shift from using guns to using missiles as their primary air-to-air armament. This made machine guns largely ineffective in defending against attacking fighter aircraft. As a result, patrol aircraft abandoned forward-firing guns and defensive turrets. Aviators and began to rely on LWTs and depth charges to attack submarines and on the speed and electronic warfare equipment of their aircraft to defend themselves from hostile fighters.

Bombs

Soon after aircraft began flying ASW patrols during World War I, they also began carrying bombs. For the next four decades, these weapons, along with depth charges, served as the primary airborne ASW weapons. Before we discuss the employment of bombs in ASW flying, let us turn our attention to how bombs cause damage and the threat they pose to the aircraft that drop them.

A bomb is composed of three primary components. The exterior portion of the bomb is a strong, frangible case that provides structural integrity. The interior of the weapon is filled with explosive material. The third portion of the weapon is an arming and fusing assembly, which keeps the weapon safe until ready to employ and triggers the explosive material at the right time.

When a bomb detonates, it causes damage to a target through two primary mechanisms. The first is a pressure pulse created through the conversion of stored chemical energy to kinetic energy during the detonation. This chemical reaction forms a shockwave, which fractures the structural casing surrounding the explosive and causes a mass of metallic or composite particles to travel outwards from the exploding weapon at high speed. The shock damage caused by the pressure pulse and impact damage caused by the expanding cloud of fragments combine to damage or destroy the target.

The cloud of fragments moving outwards from a detonating bomb poses a threat not only to the target, but also to the aircraft that dropped the weapon. Bombs create what is known as a fragmentation, or "frag," pattern when they detonate. For an aircraft delivering a weapon at low altitude, this frag pattern is potentially deadly. Aviators and weapons developers must work together to determine how they can deliver a weapon accurately while avoiding the high-speed frag pattern that results once the weapon detonates.

The earliest bombs employed during World War I were not armed with specialized fuses for ASW attacks. At first, this did not pose great difficulty to patrol crews. Early aircraft were limited in their ability to carry heavy payloads, which meant that whatever bombs carried aloft were light and had very small frag patterns. This was fortuitous for early ASW aviators.

Because early airplanes lacked any sort of bombsight or aiming equipment, horizontal bombing was extremely inaccurate. In order to attack a small target like a submarine, aviators were forced to deliver their bombs flying horizontally or in a slight dive from very low level. Flying low while bombing was not a concern because the weapons were too small to generate a frag pattern large enough to threaten the airplane during the attack.

The limited power of early bombs and the rudimentary targeting equipment meant that it was very difficult for ASW aircraft to conduct an effective attack. In order to cause sufficient damage to rupture the pressure hull of a submarine, early weapons had to physically hit the boat. As a result, a successful attack was more a matter of luck than it was due to the skills of the aviators.

During the interwar period, aviation and weapons technology advanced a great deal. First, aircraft engines became much more powerful, allowing airplanes to carry much heavier weapons payloads and reach much higher speeds. Heavier bombs meant more explosive power. On one hand, this was beneficial, because it increased the chance that a hit or a near-miss would cause damage to the hull of a submarine. However, heavier weapons also resulted in much larger frag patterns. Because targeting equipment had not advanced a great deal, ASW aircraft in the interwar period were still forced to deliver at very low altitude. This combination of heavier bombs and low-altitude delivery meant aircraft in the interwar period were highly susceptible to frag-pattern damage.

To meet this challenge, various nations began to develop specialized bombs for ASW flying. First, the physical design of these bombs was modified, with the new weapons boasting light outer casings that caused less fragmentation damage and smaller frag patterns. A larger explosive payload meant more blast damage and allowed a weapon to detonate further from a submarine and still cause sufficient overpressure to damage the hull. Second, specialized fuses were built that were tailored for delivery against a submarine. These were dual-action fuses that would detonate immediately if the weapon hit a surfaced submarine or would detonate after a slight delay if the bomb entered the water.[1]

Still, these weapons were limited in their ability to cause damage to submarines. A typical British 500-pound ASW bomb used during the early months of World War II would have to detonate within 8 feet of a submarine hull to cause sufficient overpressure to rupture the hull. Aviators of the period were still limited to delivering during slight dives or level flight at very low altitude and judging their time or release "by eye." In order to increase the chance that at least one weapon detonated close to the submarine, crews would often drop multiple weapons in quick succession. This type of pattern was known as a "stick," and resulted in a line of bombs spaced at intervals as small as several dozen feet and as large a few hundred.[2] Because these weapons were quite powerful and the aircraft were at very low altitude, aviators relied on fuses to delay detonation until the aircraft was outside the frag pattern. Unfortunately, faulty fuses proved deadly to many ASW crews during World War II, who were caught in the frag pattern when the fuse-delay mechanisms failed.[3] As the war continued, ASW forces slowly traded their bombs for specialized depth charges. We will discuss this type of weapon in the next section.

Depth Charges

During World War II, aviators and operations analysts were frustrated by the difficulty of causing enough damage to sink submarines outright. ASW aircraft of the era were much faster than their predecessors, but submarines crews could generally sight the approaching aircraft and initiate a crash dive. By the time the attacking aircraft arrived overhead and dropped their bombs, the submarine had usually disappeared beneath the waves and was deep enough that dual-action fused bombs stood little change of causing damage.

This predicament caused aviators to adopt a weapon that had been previously used only by surface ships, namely, the depth charge. Unlike ASW bombs, depth charges had very light outer casings that were meant only to provide structural integrity, not to cause fragmentation damage. Because more weight could be devoted to explosives, depth charges packed a much heavier punch than ASW bombs and could rupture a submarine hull from a greater distance. Additionally, depth charges came equipped with hydrostatic fuses, which delayed detonation until the weapon had reached a certain depth. This allowed aviators to tailor the depth of the explosion to what was most effective given operational conditions.[4]

Weaponeering for depth charges did cause several additional challenges for aviators. Aiming a bomb required aviators to solve a two-dimensional geometry problem. A bomb hit a particular place on the ocean surface and detonated a very short time later. A depth charge, on the other hand, required aviators to solve a target motion analysis problem in four dimensions: position, depth, and time. For depth charges to be effective, they had to be set at a particular depth where the submarine was likely to be when the weapon detonated. In order to attack effectively the aviators had to detect the submarine, watch it submerge, and then rapidly interpolate where they should aim the depth charges to allow them to enter the water, sink to the same depth as the submarine, and detonate close enough to cause damage. Clearly, this was a very challenging task.

Just as with ASW bombs, aviators tended to deliver depth charges in "sticks" of several weapons at a time. This increased the chance that at least one of the weapons would detonate near the target. Despite the challenges, aircrews proved quite adept at delivering their weapons at the right location as their targets submerged. What proved to be difficult early in World War II was choosing the correct depth for the depth charges to detonate.

Operations analysis revealed that most Allied depth charges used early in the conflict were detonated too deep. Specialized shallow-water fuses were developed, which measurably increased the effectiveness and lethality of airborne ASW attacks. Operations analysts also helped aviators pick the best location relative to a recently submerged boat where they should drop their weapons. These efforts proved very successful at increasing the rate of successful attacks.[5]

With the end of World War II and the advent of the Cold War, torpedoes replaced conventional depth charges as the primary method of attacking submarines. However, the advent of nuclear weapons introduced the possibility of compact and extremely powerful new weapons to employ against submarines. The introduction of nuclear-armed SLBMs and cruise missiles meant that submarines became strategic targets. The necessity to destroy a hostile SSBN or SSGN before it could launch its weapons led to the development of nuclear depth charges.

Many ASW aircraft were equipped with nuclear depth charges during the Cold War.[6] The enormous explosive power of these weapons meant that a nearby submarine could be sunk or disabled even if its location could not be precisely fixed. A nuclear depth charge dropped within several miles of a submarine was likely to cause heavy damage, if not sink the target outright.

Beyond the overpressure caused by a nuclear explosion, these so-called "special weapons" caused other serious side effects. Testing revealed that underwater nuclear explosions caused enormous disruptions in the water column. High levels of reverberation and a disruption of the area frequently prevented the use of sonar for some time after the explosion. These conditions would severely hamper the attacking aircraft from evaluating whether the target was destroyed or remained operational.[7]

Thankfully, the Cold War ended without nuclear depth charges or any other nuclear weapons being used in anger. It is most likely that nuclear capabilities have been removed from all ASW aircraft worldwide, although it is difficult to verify whether Russian ASW aircraft ever carried nuclear depth charges during operational missions or if these aircraft are still wired for such weapons.

Air-to-Surface Missiles

A surfaced submarine has always proven an inviting target, especially for aircraft armed with powerful forward-firing weapons. During World War II, Allied pilots flying light and medium bombers attacked U-boats using unguided rockets. These weapons were useful not only as anti-personnel weapons but also boasted enough firepower to penetrate submarine pressure hulls and prevent the boats from submerging.

As miniaturized electronics and radar came of age after World War II, aircraft gained the capability to fire guided missiles. Early ballistic missile submarines and cruise missile submarines could not fire their weapons underwater and were obliged to surface while they launched. Some SSGNs and SSGs had to not only surface to fire but also remain exposed while they guided their weapons to impact. While on the surface, these boats were very vulnerable to attack.

This led many navies to equip their ASW aircraft with guided air-to-surface missiles. Some of these weapons were optically guided and some were optimized to detect the relatively small radar cross section of a submarine and guide to impact.[8] The large size of anti-ship missile warheads made it likely that a successful attack would cause serious structural damage and prevent the boat from submerging.

Modern ballistic missile and guided missile submarines are capable of submerged launch. Equipped with modern weapons and propulsion systems, there is no compelling reason for a boat to fully surface during wartime. As a result, patrol aircraft have very little opportunity to attack a submarine in the open ocean using air-to-surface missiles and few ASW aircraft routinely fly with this type of weapon. That is not to say that modern patrol aircraft do not carry missiles. On the contrary, they frequently are equipped to carry ASCMs, but usually only fly with these weapons when they are engaged in ASUW operations or combat patrols.

Mines

Once early aircraft began carrying bombs, the introduction of air-dropped mines was not far off. The speed and range of aircraft meant that they could quickly reach an operating area and lay a minefield in short order. Their speed and maneuverability provided a degree of protection from AAA and fighter aircraft and heavily armed patrol aircraft could protect themselves using defensive guns.

Minefields are laid in different areas, depending on the objectives of the ASW force. Offensive mining may be conducted near adversary ports to damage boats as they come and go, to force boats through narrow choke points, or to force the enemy to expend time and energy sweeping the field. Laying offensive minefields near enemy ports means that in all likelihood there will be significant air defenses for the mining aircraft to contend with. Because most modern ASW aircraft are derivatives of airliners, they are lightly armed and intolerant of damage. Thus, offensive mining in the modern era is generally left to strike fighters or bombers.

Minefields may also be laid in areas further from adversary shores. The geography of certain areas may mean that submarines must pass through shallow choke points to reach their patrol areas. These choke points are natural locations for offensive minefields.

Last, an ASW force may lay defensive minefields to protect ports or patrol areas where friendly units are operating. For example, aircraft may lay a minefield in between the adversary and a patrol area where a friendly SSBN is operating. This requires a hostile SSN to transverse a minefield before reaching the operating area to hunt for the SSBN.

Mines can be divided into broad categories based on where they are deployed, how they are moored, and how they are activated. Generally, mines are designed to be deployed in a certain depth of water based on their size and design features. In terms of mooring, mines may be free floating, moored to the bottom but floating at a shallower depth, or stationary on the bottom. Most mines deployed by aircraft today are bottom mines and are derived from general-purpose bombs.

Mines are fused in several different ways. Acoustic fuses arm the weapon when a sonar array detects the sound of a ship or the acoustic signature of a particular class of warship or submarine. Magnetic fuses are set off by the flux in the earth's magnetic field caused by the passage of a ship's hull. Influence fuses are activated by the pressure differential caused by water that is displaced by a ship moving nearby.

Some navies have deployed mines that are more akin to unattended weapons systems. Encapsulated torpedoes are specialized mines that consist of a sensor package, a launch mechanism, and a LWT. When the sensor package detects the characteristic tonals of a target submarines, it activates the launch mechanism, which ejects the torpedo from the mine casing. The torpedo swims out after the target and begins a search pattern to acquire and home on the adversary boat.[9]

The effectiveness of bottom mines is driven by the size of their explosive payload and the depth of the water where they are laid. Smaller mines are effective in shallow water but if employed at a greater depth would lack sufficient explosive power to damage a ship or submarine transiting on the surface. Therefore, smaller bottom mines are used in shallow water while larger bottom mines are used in deeper water. Bottom mines are typically laid in transit lanes near enemy ports or in shallow choke points that submarines must pass to reach patrol areas.

Torpedoes

For the last 60 years, the primary weapon used by ASW aircraft to attack submarines has been the LWT. During World War II, engineers developed a simple torpedo that used a passive hydrophone to home on the cavitation sounds produced by a submarine evading at shallow depth. The introduction of an effective air-dropped LWT was fortuitous, because the adoption of the snorkel and introduction of nuclear power several years later rendered ASW bombs and depth charges far less effective. With boats remaining underwater for the vast majority of their time at sea, aviators needed a weapon that was effective against a submerged target.

Torpedoes are guided either by a passive or active sonar seeker head. When a LWT is dropped, the weapon enters the water, activates, and swims to a pre-set activation depth. Here, the torpedo begins a search pattern using passive sonar, active sonar, or a combination of both to detect the target. Because LWTs can run a search pattern, they can be deployed with a certain amount of uncertainty as to the exact location of the target submarine. This is important, because tracking a boat underwater means there is generally a measure of ambiguity and uncertainty as to the exact location of the submarine.

The earliest LWTs relied primarily on passive sonar to detect their targets. This was due to the fact that World War II-era submarines and those in the immediate post-war years were generally quite noisy and cavitated readily.[10] As navies worked to decrease the radiated noise levels of their boats, LWTs began to rely more on active sonar guidance. The seeker head of these weapons could detect reflections of sound energy off of the hull, screws, or wake of the target submarine.

However, the use of active sonar was not without drawbacks. In shallow water where the submarine is close to the bottom, the seeker head may be fooled by the presence of wrecks, pinnacles, bottom features, or clutter in the water. The introduction of SSKs and midget submarines (SSMs) that could sit on the ocean bottom created a new challenge for aviators attacking with LWTs. Engineers were forced to develop sophisticated signal-processing techniques to reject the false returns caused by the ocean bottom and to break the target out from background clutter.

Because submarines were usually built with relatively fragile hulls, early torpedo designs were equipped with bulk charge warheads. However, during the 1970s, the Soviet Union began to build submarines with more durable double hulls to improve survivability and reduce the effectiveness of Western weapons. In order to rupture the hull, a LWT warhead would have to punch through the hydrodynamic outer hull, a void area between the outer hull and pressure hull, and finally penetrate the pressure hull. The inability of Western LWTs to accomplish this task forced ASW forces to field new LWTs equipped with shaped-charge warheads.[11]

A shaped charge consists of explosive material and a casing structure built in a special configuration. When the warhead detonates, the explosive force is channeled to produce a high-speed jet of molten metal in a focused direction. A LWT with a shaped charge aims to detonate and punch through the outer hull, void space, and inner hull, creating spall damage or causing flooding in the affected compartment.

Because LWTs are often launched with uncertainty about the target's location, they must be fast enough to run down and catch a fleeing submarine. Generally, engineers have designed weapons to have a maximum speed at least 50 percent greater than the maximum speed of a target submarine. Additionally, the torpedo must be able to dive deep enough to pursue the evading boat.

LWT warhead fusing can be accomplished in several ways. Contact fuses detonate the warhead when the torpedo strikes the hull of the target. Magnetic fuses are used to explode the warhead when the torpedo passes in very close proximity to the hull but does not actually make contact during the approach. Should a torpedo miss, most seeker heads contain guidance algorithms that force the weapon back into a search mode in an attempt to reacquire the target and make another approach.

Submarine Self-Defense Weapons and Countermeasures

So far, we have focused our discussion on the offensive weapons used by ASW aircraft. Now, we will turn our attention to those weapons that submarines have historically used to protect themselves from aerial attack. We will also review the various defensive countermeasures that submariners have at their disposal.

Anti-Aircraft Artillery

Because early submarines spent a majority of the time on the surface, they frequently were equipped with deck guns. As aircraft began flying ASW patrols during World War I, submarines began carrying deck-mounted AAA to defend themselves. Because of size constraints and stability concerns, submarine mounted AAA was small in caliber and limited in range. However, it did pose a serious threat to attacking aircraft, especially given the flimsy construction of early patrol aircraft.

During World War I it was relatively unusual for submarines to have to defend themselves because attacking aircraft were so slow. Lookouts would usually spot an attacking aircraft and alert their compatriots. Given that ocean-going submarines could submerge in a minute and coastal submarines could submerge in as little as 30 seconds, it was much safer and more effective to get underwater and evade rather than try to fight off an attacking aircraft with gunfire.

By World War II, however, AAA became more relevant and was used far more frequently. The higher speed of ASW aircraft during this conflict meant that it was far more likely for aviators to get overhead and attack before the boat could submerge. Submariners would generally submerge if they sighted an aircraft early enough, but if the sighting was late, crews would remain surfaced to try to shoot down the attacking aircraft rather than offer the aviators an unopposed approach and good chance of accurately delivering a lethal attack against the partially-submerged boat.

After the submerged revolution, submarines stopped spending large amounts of time on the surface. This removed any necessity to mount AAA. Conventional boats removed their AAA mounts to save weight, reduce drag, and cut down on flow noise. Nuclear submarines never had any reason to mount AAA. By the 1950s, the era of deck-mounted guns aboard submarines had passed. These weapons are unlikely to return.

Man-Portable Air Defense Systems

During the Cold War, armies had a great need to protect fast-moving maneuver forces from attacks by aircraft and helicopters. The proliferation of miniaturized electronics components led to the construction of SAMs small enough to be carried by solders on the battlefield. This led to the introduction of so-called Man Portable Air Defense Systems (MANPADS).

The small size of MANPADS means that these weapons can be carried aboard submarines and used while the boats are on the surface.[12] Most MANPADS are equipped with IR guidance systems, which detect the heat produced by aircraft engines and high-temperature components and use these radiation sources to home to impact. MANPADS are generally cued by the naked eye, aligned for launch by the operator, and do not require any guidance from the operator once they are fired.

MANPADS are quite fast and maneuverable, but the small size of their rocket motors means that they are only effective over ranges of a few miles. However, modern ASW aircraft are intolerant of damage and low in maneuverability, so aviators must respect these weapons and remain wary when they spot a submarine on the surface with personnel outside the hull.

MANPADS must be fired by an operator in the sail of the submarine or on the deck. This means that submarines must go through a lengthy process of surfacing and then put personnel above. While a patrol aircraft must stand off from a submarine armed with MANPADS, the submarine has lost any advantage of stealth or mobility. The aviators can still attack using stand-off missiles or call in strike aircraft, which are much less susceptible to MANPADS due to their speed, maneuverability, and defensive systems.

MANPADS, however, are very useful while a crew is abandoning ship or scuttling a submarine. By holding aircraft at bay, they allow a crew to complete an evacuation or buy the time necessary to scuttle the boat and prevent its capture by the enemy. Despite their limited size and range, MANPADS remain effective weapons and aviators are wise to remain wary of them.

Submarine-Launched Surface-to-Air Missiles

Under certain conditions, submarine passive sonars are capable of detecting low-flying ASW aircraft as they overfly the boat at close range.[13] Additionally, submariners can detect an aircraft nearby using ESM equipment or visual observations with a periscope. For the last several decades, these detections served only to alert the submarine crew that an aircraft was overhead. Without a way to accurately plot the location of the submarine, there was no chance of striking back against the aircraft.

During the last 30 years, engineers have worked in fits and starts to equip submarines with the capability to attack aircraft flying overhead while remaining submerged.[14] Recent years have witnessed defense firms developing submarine-launched missiles that can be fired underwater, rise to the surface, and attack aircraft overhead. These weapons use input from sonar and periscopes to plot the location of an aircraft. The missiles are released by the submarine, rise to the surface, and ignite their rocket motors. Once airborne, they are guided using optical guidance from the submarine or through the use of an IR seeker.

While these submarine-launched SAMs are not yet operationally deployed, the technology is advancing rapidly. These weapons are most effective against slow-moving ASW helicopters that are loitering above a boat or hovering to deploy a dipping sonar. A stationary helicopter is a much simpler target to detect and attack than a fast-moving ASW aircraft. As a result, these weapons are more threatening to helicopters than to fixed-wing aircraft. However, because of their fragile construction, modern ASW aircraft remain vulnerable to these weapons and aircrew should respect this threat by remaining at a safe distance from a target submarine or operating at altitudes outside their engagement envelopes.

Submarine Evasion Techniques and Countermeasures

To best employ their weapons, ASW aviators should understand the strategies that submariners will use to degrade an attack. Because of their vulnerability and lack of defensive systems, submarines are obliged to take the presence of any weapon in the water very seriously. Even small explosions close aboard can have dire consequences to a submerged boat due to secondary effects from flooding, fire, or shock damage.

When under attack, submariners employ two primary methods to escape from harm. The first method is using maneuvers to reduce the chance that a torpedo seeker will acquire the submarine in the first place. The second method is using a combination of maneuvers by the boat and the employment of countermeasures equipment to degrade a seeker that has acquired or is guiding. We will discuss both strategies in detail.

A submarine crew that detects a torpedo in the water will likely first maneuver to escape the field of view of the seeker head. Boats will probably use a combination of speed and course changes to attempt to exit the area where the torpedo's seeker head is searching. Additionally, submariners will attempt to put obstacles between their boat and the torpedo that reduce the likelihood that the seeker can detect the submarine. For example, a boat in deep water may maneuver to the opposite side of the thermocline, relying on the effects of refraction and reflection to turn active sound energy emitted by the torpedo seeker.

If a torpedo is dropped at close range and a submarine crew feels they cannot evade the coverage of the seeker, they will probably then focus on defeating the seeker itself. This is done using maneuver, countermeasures equipment, or a combination of both. Submarines can utilize two strategies to defeat a torpedo seeker through maneuver. The first is taking advantage of the susceptibility of torpedoes to cease homing against stationary targets while the second is turning the boat so as to present the smallest active sonar return possible.

Active sonar seekers are designed to reject the presence of clutter caused by the seafloor and countermeasures. To do this, a seeker looks for sound energy that has reflected off a moving object and had its frequency shifted through the Doppler effect. If acoustic energy has not been shifted in frequency, the signal processing algorithm assumes that the sound waves have reflected off a stationary object such as the seafloor or a cloud of bubbles caused by a countermeasure. The algorithm rejects this information and seeks a moving target that is more likely to be the submarine. Submariners are aware of the propensity of torpedoes to overlook stationary objects and seek moving objects. As a result, they may utilize tactics where they slow or stop dead in the water to hide amongst the clutter.

Submarines may also maneuver to present a smaller target to a torpedo's active seeker head. The target strength (TS) of sonar returns reflected by different parts of a submarine is not uniform. Rather, the amount of energy returned varies in azimuth, with more energy reflecting off the sides of a submarine and smaller amounts reflecting when viewing the submarine head-on or stern-on. To present as little acoustic energy as possible to the incoming torpedo, a submarine may turn head-on or stern-on to minimize the TS observed by the seeker head.

Submarines are also equipped with a variety of countermeasures used to directly degrade torpedo seekers. These systems are either mobile or static. Mobile countermeasures tend to be more sophisticated and advanced, using several techniques to decoy an incoming weapon away from the submarine. Static countermeasures, on the other hand, are generally simpler and less expensive, allowing many to be deployed during combat.

The acoustic decoy is one type of countermeasure. This system uses a signal generator and sound transmitter to emit noises and lure passive seeker heads away from the actual sound of the submarine. These decoys may either emit loud broadband noise or emit specific narrowband tonal frequencies that mimic the signature of the target submarine.

A simpler and less expensive version of the acoustic decoy is the general-purpose noisemaker. These devices generate noise through one of two mechanisms. The first is the release of bubbles, which produce high levels of broadband noise. The second is the use of mechanical equipment, which may produce precise narrowband tonals or more diffuse broadband noise.

Bubble generators are static countermeasures that use a chemical reaction to create a cloud of air bubbles. These bubble clouds serve two purposes. The first is to generate strong active sonar returns, hiding the returns from the submarine with returns from the bubble cloud. The second purpose of the bubble generator is to produce high levels of broadband noise, drowning out the actual signature of the submarine and degrading passive seeker heads.

Another type of countermeasure is an echo repeater. These systems receive incoming sonar pulses and use transceivers to repeat back sound energy to the torpedo seeker head. The return signal may be shifted in frequency to present Doppler to the torpedo and register as a valid target rather than as stationary clutter. This causes the seeker head to home on the echo repeater, rather than on the sound energy reflected off the actual target submarine itself.

Frequently, several countermeasures techniques may be combined together. For example, various submarines classes carry mobile countermeasures that feature high fidelity acoustic decoy capabilities and echo repeaters to degrade active sonar and passive sonar.[15] When under attack, the submarine crew is likely to deploy these advanced mobile decoys along with a larger number of static acoustic decoys and noise-makers.

Notes

1. For details of purpose-built ASW bombs developed by the British, see Alfred Price, *Aircraft Versus Submarine in Two World Wars* (Barnsley, South Yorkshire: Pen and Sword Books, 2004), pp. 34–5.
2. Ibid., p. 35.
3. In what was probably the first attack by an aircraft on a submarine in World War II, an RAF Anson patrol plane dropped bombs on a submerging submarine off the west coast of Scotland on 5 September 1939. The bombs skipped off the surface and detonated, showering the Anson with fragments and forcing the crew to ditch. The crew was rescued afterward and were happy to find out their attack had done no damage, given that the target was actually a British submarine, HMS *Seahorse* (98S). Ibid., pp. 39–40.
4. Ibid., p. 48.
5. Stephen Budiansky, *Blackett's War: The Men who Defeated the Nazi U-boats and Brought Science to the Art of Warfare* (New York: Random House, 2013), pp. 141–5.
6. Norman Polmar and Edward Whitman, *Hunters and Killers Volume 2: Anti-Submarine Warfare from 1943* (Annapolis, MD: Naval Institute Press, 2016), pp. 171–2.
7. McMillan de le Houssaye, Johnson, "OPERATION DOMINIC, Shot SWORD FISH, Project Officer's Report – Project 1.3b, Effects of an Underwater Nuclear Explosion on Hydroacoustic Systems" (San Diego: U.S. Navy Electronics Laboratory, 1 September 1985).
8. The American Harpoon family of ASCMs was originally developed to be carried by patrol aircraft to attack surfaced Soviet SSGNs. The weapon was later adapted to launch by ships, submarines, and aircraft to attack surface ships.
9. For details on the American Mk 60 CAPTOR (enCAPsulated TORpedo) mine, see Polmar and Whitman, *Hunters and Killers Volume 2*, pp. 113–14.
10. The first airborne homing LWT was the American Mk 24. This torpedo, nicknamed "Fido," was introduced in March 1943. Mk 24s scored kills in roughly 18 percent of the engagements when they were employed, a significant increase over the approximately 9 percent kill rate of attacks conducted by aircraft using bombs or depth charges. Norman Polmar and Edward Whitman, *Hunters and Killers Volume 1: Anti-Submarine Warfare from 1776 to 1943* (Annapolis, MD: Naval Institute Press, 2015), pp. 166–8.
11. Polmar and Whitman, *Hunters and Killers Volume 2*, pp. 94–5.
12. Soviet Project 949/TYPHOON class SSBNs and Project 636/877/KILO class SSKs were built with a position in the after portion of the sail for a sailor to handle and fire MANPADS. Each boat has stowage space for eight missiles. Norman Polmar and K.J. Moore, *Cold War Submarines: The Design and Construction of U.S. and Soviet Submarines* (Washington, D.C.: Potomac Books, 2004), p. 302.
13. For a discussion of Soviet submarines counter-detecting low-flying NATO patrol aircraft, see Tony Blackman, *Nimrod: Rise and Fall* (London: Grub Street Publishing, 2013), p. 75.

14. Both the British and Americans experimented during the 1980s with equipping submarines with the capability to fire Blowpipe and Sidewinder missiles from broached or submerged boats. Neither of these capabilities was deployed operationally. Polmar and Moore, *Cold War Submarines*, p. 302.

15. Consider the example of the U.S. Navy Mk 70 MObile Submarine Simulator (MOSS) mobile decoys, carried aboard American SSBNs during the Cold War. Edward Monroe Jones and Shawn S. Roderick, *Submarine Torpedo Tactics: An American History* (Jefferson, N.C.: McFarland, 2014), p. 179. Soviet submarines employed similar decoys, such as the MG-74 Korund mobile decoy carried by Project 949/TYPHOON class SSBNs, Project 945/SIERRA class SSNs, and Project 917/AKULA class SSNs. Polmar and Moore, *Cold War Submarines*, p. 284.

Chapter 7

Airborne ASW Sensors

For ASW aviators to properly employ their sensors, they should understand how these tools function and the basic physical phenomena that make submarine detection possible. Sensor theory is well documented in the academic world. The mechanisms of how sonar arrays, radars, and MAD equipment function can all be deduced from first principles. It is the unique characteristics and capabilities of these sensors that are sensitive, not the general manner in which they work.

The following chapter is a basic introduction to airborne ASW sensors. It is meant to serve as a primer and to stimulate interest. Entire volumes have been written on sonar, radar, and signal processing alone. Aviators can find much more in-depth and specialized discussion in other works.

Visual

Searching for submarines with the naked eye is the oldest form of airborne ASW sensing. Despite the changes in technology and tactics, detecting submarines visually is still relevant today. If the water is sufficiently clear, submarines can be observed underwater while they are at periscope depth or running shallow. This is unusual, because it requires very clear water, but is certainly possible, especially in some parts of the tropics and in the Adriatic Sea. This is due to these waters having very low levels of turbidity, the measure of clarity of water.

A more common visual detection method is spotting the wake caused by a surfaced submarine, an exposed periscope, or an extended mast. Wakes are bigger than the objects that cause them and may persist for some time. This means that aviators typically spot a wake before they spot the object that caused it. A periscope moving through the water causes a distinct wake, called a "feather." The intensity of the feather grows with the speed of the submarine. It is more easily observed in a low sea state. A feather caused by a submarine moving slowly in a wind-driven, choppy sea is much harder to spot than one cause by a boat moving quickly in a low sea state.

The orientation of the wake or feather either helps or hinders the aviators. Wake detection usually works best at low grazing angles. This drives aircraft that are searching visually to fly at altitudes of a few thousand feet or lower. Visibility and the angle of the sun also have a large effect on visual searches. Observers are better at spotting an object in the water with the sun at their backs than while they are looking up-sun. Haze hurts visual search, especially when the sun is at low angles during the early morning or late afternoon.

Modern ASW aircraft are frequently equipped with EO cameras and magnification equipment to observe and identify targets at long range. They may also have IR cameras that allow aviators to observe objects at night and in low light. EO and IR cameras achieve good long-range magnification at the price of a very small field of view. In practice, their limited scan rates and fields of view makes them identification tools rather than search sensors. They are most useful when an operator detects an object visually or with radar and then slews the long-range camera for identification.

In theory, it is possible to visually detect bioluminescence created by a submarine or detect with an IR sensor differences in surface water temperature caused by submarines disturbing the thermocline below. In practice, these methods are usually ineffective visual search methods. We will discuss them later in the chapter during the sections on wake detection and bioluminescence.

Radar

During World War II, Allied scientists developed radar systems small enough and powerful enough to be carried on bombers and ASW aircraft. These early airborne radars were quickly pressed into service detecting surfaced submarines. They could detect submarines at longer ranges than visual search methods and at night when visual detection had previously been impractical. The greatest challenge was discriminating between energy that reflected off a submarine and the clutter created by waves and the ocean. This caused engineers to focus on more powerful sets with greater resolution.

By the end of World War II, the introduction of the snorkel meant airborne radars had to pick out a mast a few feet tall rather than the entire submarine hull and conning tower. This radically reduced the effectiveness of airborne ASW radars. Engineers focused on using short pulse-width, high pulse-repetition frequency waveforms to separate radar returns caused by a snorkel or periscope from the clutter caused by ocean waves. By the

1960s, NATO air arms had developed radars capable of detecting snorkels and periscopes. These became very useful sensors for forcing submarines to remain submerged or detecting them while they were snorkeling to recharge their batteries.

Radar uses pulses of electromagnetic energy to reflect off parts of a submarine above the ocean surface. It is an overt sensor. Like active sonar, a submarine can detect radar at a far longer distance than the radar can detect the submarine. This is not necessarily a negative for an ASW aircraft. Periscope-detection radars have distinctive signatures and are easily identified by ESM systems aboard submarines. Since submariners are very conscious of the threat posed by ASW aircraft, detecting an airborne ASW radar nearby is a powerful deterrent.

Radar flooding has traditionally been a useful tactic against submarines, especially against conventional boats that must expose their snorkels periodically to recharge their batteries and ventilate the atmosphere.[1] A conventional submarine being held under by radar flooding must stay at low speed to husband its battery charge. This robs the boat of its mobility. Conventional or nuclear submarines that expose their periscope during an attack are also held at risk if a radar-equipped ASW aircraft is nearby.

Like visual search, airborne radars tend to perform best at low to medium grazing angles. Some modern ASW radars make use of a technique known as inverse synthetic aperture radar (ISAR). This mode uses motion of an object in a seaway to generate an image of the target. This can be used to classify the radar contact, as a submarine snorkel will appear different on an ISAR return compared to a fishing boat.

Passive Sonar

As submarines shifted from operating primarily on the surface to operating mostly underwater in the 1950s, ASW aircraft were robbed of their principal submarine detection method, radar search. This forced aviators to shift to detecting submarines using sonar. Conventional submarines snorkeling and running diesel engines were noisy, as were the early nuclear submarines of the era. Passive sonar was a natural solution to detect and track the sound energy emitted by these boats. It remains the primary search and tracking sensor used by ASW aircraft today.

Passive sonar relies on detecting the acoustic emissions of a target submarine. It does not emit energy, making it largely a covert surveillance tool.[2] Passive sonar systems consist of a hydrophone or group of hydrophones that detect acoustic energy. Hydrophones are composed of elements known as transducers, which detect variations in sound pressure level (SPL) and generate an electric impulse. Multiple transducers are grouped together and their signals are combined for processing. The size, shape, and arrangement of the transducers are designed in such a way that the hydrophone performs best against a particular frequency range and responds best to sound arriving from a particular direction.

An individual transducer cannot register a sound wave that is smaller than itself. Since the amplitude of a sound wave is inversely proportional its frequency, this drives an upper frequency limit of the hydrophone. The size of the entire hydrophone array drives the lower limit of the frequency range, as it cannot register a wave larger in amplitude than itself. An ideal hydrophone array could consist of a large number of very small transducers and be very large overall. This would provide small transducer size to respond to high-frequency sound while making the array large enough to respond to low-frequency sound. However, building a large array with many small transducers is expensive, driving engineers to compromise.

The output of the hydrophone array is fed to an acoustic processor. The processor attempts to pick out the signal generated by the submarine from the noise in the ocean. Engineers quantify the relationship between the strength of the submarine signal and the background noise using a term known as the Signal to Noise Ratio (SNR). To maximize the range at which a submarine can be detected, the engineers try to boost the SNR by driving down the noise received by a hydrophone as low as possible.

When engineers design a hydrophone, they attempt to influence its ability to detect sound from a particular direction. They do this partially by controlling the physical arrangement of the transducers and partially through a signal-processing technique known as "beam forming." Sound energy from a submarine, the strength of which we will refer to as Source Level (SL), arrives from a particular direction while Ambient Noise (AN) arrives from all directions. By only gathering sound energy from a particular direction, the intensity of the AN drops while the submarine signal strength stays the same. This boosts the SNR, causing an effect known as "gain." This characteristic of a hydrophone to respond to sound from a particular direction is known as "directivity" and is quantified by a value known as Directivity Index (DI).

Submarines emit sound, the strength of which is generally measured one meter from the hull and expressed in decibels (dB). The powerplant, auxiliary systems, and interaction of the submarine with the ocean will produce a spectrum of acoustic energy. The intensity and particular frequencies where a boat radiates energy will vary, with some frequencies being very pronounced and noisy and others much quieter. The frequencies and intensities of this spectrum will shift depending on the speed and depth of the submarine and what equipment is operating. To better describe submarine sound signature, we will break their sound sources down into two categories: narrowband sound sources and broadband sound sources.

Narrowband sound sources cluster tightly around a particular frequency. These precise, well-defined sound sources less than 1 Hz in width are often the result of rotating machinery or electrical systems that run at a particular frequency.[3] Because these sources vary very little, they are useful to identify a particular class of submarine or even an individual boat. Narrowband sources are very well suited for signal-processing techniques. They also lend themselves well to tracking quiet targets and submarines moving at slow speeds.

Narrowband sources are usually displayed on a screen known as a Low-Frequency Analyzer-Recorder gram, or LOFARgram. A LOFARgram, usually known simply as a "gram," shows the frequency spectrum, with low-frequency sound energy on the left and high-frequency sound energy on the right. The gram moves upwards with time, allowing operators to analyze signals registered on a hydrophone and note how they change with time. By observing variations in frequency and direction displayed on the gram, aircrew can perform TMA. A notional LOFARgram may be found in **Figure 1**.

Unlike narrowband noise, broadband noise is diffuse, generally spreading over more than 1 Hz. Broadband noise is often produced by water flowing over the hull of a submarine or by propeller cavitation.[4] It is not as useful for providing target classification but is very useful for tracking fast targets, especially nuclear submarines. Aircrews frequently refer to flow noise caused by the hull moving through the ocean as "sea water across the hull" or a "swath."

Changes in how the submarine operates will cause narrowband and broadband frequencies and source levels to change. For example, a nuclear submarine moving at low speed will have its coolant pumps operating slowly, turning at a low frequency and making relatively little noise. To drive the boat faster, the reactor is adjusted to produce more energy, in turn generating more steam and driving the turbines faster. This requires more waste heat to be drawn away from the reactor. The crew will shift their coolant pumps to high speed, which will produce a higher-frequency and noisier signature. As the boat accelerates, flow noise over the hull will become more apparent and the propeller may cavitate. These actions will change the frequency and source levels of the narrowband sources produced by the coolant pumps and increase the broadband noise caused by the boat.

By analyzing acoustic data and drawing inferences about the design and construction of a particular submarine, intelligence analysts can build a description of the acoustic signature of a submarine class or even an individual boat. They will list the various narrowband sources, or tonals as they are more commonly called, and the associated strengths of each source.[5] They will also describe the various broadband sources, when they occur, and their strengths. They will describe how the tonals and broadband sources vary as the boat operates at low, medium, and high speeds.

Aviators can then identify which of these sources propagate best in the ocean and can be detected or tracked. Low-frequency noise is less attenuated by the ocean, and can travel further, but is often masked by AN. Low-frequency sources may be useful for initial detection but are not always well suited for tracking. Unique sound sources which aren't masked by AN or are higher in frequency are often the best choices to focus on to identify and track the submarine. When planning a mission, aviators identify the particular "frequencies of interest" they will focus on while searching for and tracking their target.

When acoustic energy is detected on a hydrophone, it is converted into electrical impulses. These electrical signals are grafted onto a radio carrier frequency and transmitted from sonobuoys to the ASW aircraft. Aboard the aircraft, the signals are stripped from the carrier wave and ported into the acoustic processor. The acoustic processor uses signal-processing techniques to separate the acoustic energy emitted by the submarine from the background noise of the ocean. The resulting information is displayed on screens for visual analysis and output for aural analysis by acoustic operators onboard the aircraft.

Passive sonobuoys provide aviators with a bearing to a sound source and a readout of the frequency of that signal. The bearing will have a particular amount of error based on the accuracy of the sensor. The frequency displayed will be that received by the hydrophone array, not the actual frequency produced by the sound source on the submarine. This is because as sound moves through the water it is shifted in frequency by the relative motion between the sensor and the boat and variations in the water column.

In analyzing sensor performance, aviators and engineers try to quantify the minimum level of signal strength that can be detected by a human operator. This term is known as "Recognition Differential," abbreviated as RD. For the sake of convention, RD is defined such that a signal is sufficiently strong that it can be detected by an un-alerted observer 50 percent of the time.

RD is influenced by the combination of the sensor, processing, display, and analysis process. A higher-definition LOFARgram display will allow an operator to pick out a weaker signal, effectively raising the RD. A fatigued operator flying in the middle of the night will perform less effectively than a fresh, well-rested operator and in turn, the sleepy operator's poor performance will cause RD to suffer. Any change in the combination of sensors, processing, display, or operator performance will affect the RD realized by the human-machine team aboard the ASW aircraft.

As sound waves move through the ocean they lose intensity as they interact with the medium of travel. This energy dissipation is known as Transmission Loss, abbreviated as TL. Sound moving in the ocean loses energy due to scattering, spreading, and absorption. The properties of the water column govern how fast spreading and absorption occur. Scattering is due to sound waves hitting an interface, such as water with different properties, marine life and particulate matter in the water, or the ocean bottom. We will discuss the properties that govern these phenomena in more detail in the Chapter 8.

If the properties of the water column and bathymetry are known, the rate that TL occurs can be calculated. The rate of TL can then be plotted on a curve. In **Figure 2**, a typical TL curve is shown. It should be noted that TL is not constant. This is due to environmental effects we'll discuss in further detail in Chapter 8 and Chapter 9.

The final measure we must consider before we can fully describe passive sonar sensor performance is the Noise Level (NL) experienced by the sonobuoy. NL is an expression of the background noise received by the hydrophone. It is the product of Self Noise (SN), the sound generated by the interaction of the sonobuoy with the environment, and AN, the manmade and environmental background noise in the ocean.

SN is a measurement of sound around a hydrophone. In the case of passive sonobuoys, it is caused by two sources.[6] The first is flow noise of water moving around the hydrophone. The second is due to a phenomenon known as "cable strumming." Sonobuoy hydrophones are suspended deep in the water by a cable that connects them to a flotation bag and radio antenna on the surface. As water flows around the tether between the hydrophone assembly and the flotation bag, vortices form that make the cable vibrate. This "strumming" causes noise around the hydrophone.[7]

AN is primarily a result of distant shipping noise, biological noise, and noise created by wind and wave action. Shipping noise and biological noise vary both temporally and spatially. We will discuss AN and the factors that cause it in much greater detail in Chapter 8.

Now that we've discussed each of the factors that impact passive sonar performance, we can combine them together to form the passive sonar equation. To describe the amount of energy remaining at a particular distance from a source, we use the term Signal Excess (SE). Below, we can find the passive sonar equation.

$$SE = SL - TL - NL + DI - RD$$

Several comments should be made about the passive sonar equation. First, the terms and relationships hold for a particular frequency of interest. As we mentioned before, the ocean will attenuate various frequencies at different rates. AN will be stronger or weaker at different frequencies. Submarines emit different SL values at particular frequencies. Therefore, it is important to recognize that the terms of the passive sonar equation hold only for a particular sound source, with its corresponding frequency and source level, and in a particular water column, with its unique sound speed profile (SSP), bathymetric features, and AN value.

As we mentioned above, if we can characterize the water column and bathymetry, we can accurately predict TL. Since TL varies with range, we can substitute this function into the passive sonar equation and solve for SE as a function of range. This is very important for ASW operations, because it allows aviators to predict how far from a sensor they can detect a submarine.

Calculating TL curves has typically been done with computers, and for most of the last several decades, this could only be done at operations centers ashore. Aviators would use historic SSP and AN values from environmental databases and generate TL curves before they went flying. However, they needed a way to quantify how much more or less likely they were to detect their target if the ocean environment was different when they arrived on station.

To help adjust to varying environmental factors, aviators often used a concept known as Figure of Merit (FOM). To calculate FOM, they would set SE to zero and rearrange the values of the passive sonar equation. The results are shown below.

$$FOM = SL - NL - TL - RD + DI$$

FOM evolved from early sonar experiments, where scientists and engineers were trying to quantify how well the sonar was performing. A particular FOM would describe the ideal performance and detection capabilities for the sonar in question in that particular water column against a particular source. Aviators would analyze and plot a predicted FOM given their best estimates of the target, the SSP, and the bathymetry in the search area. Once they arrived on station, they would sample the water column and determine differences in the AN and SSP. Using the original FOM and TL curve, they would then compensate for these values and determine how much longer or shorter their detection ranges would be. Depending on how high or low the various parameters were, the aviators could determine if the SE was above or below the expected value for their FOM calculation. We will discuss these variations in parameters further in Chapter 9.

As computing power improved, it became possible to rapidly calculate TL using environmental data measured by sensors in the water. Ships and submarines began using integrated environmental modeling tactical decision aids, which would calculate TL curves in real time using in-situ environmental data. While the computing power required to calculate TL curves is not high, ASW aircraft have lagged ships and submarines in adopting integrated tactical decision aids. This is due to the fact that computers fast enough and small enough to fly aboard patrol aircraft were developed just as the Cold War ended and funding for ASW forces was drastically reduced.

Active Sonar

Unlike passive sonar, which listens for acoustic energy emitted by the submarine, active sonar systems radiate energy of their own. Their sound pulses travel through the water, reflect off the submarine, and travel back to a receiver. While passive sonar gives only bearing and frequency information, active sonar yields bearing, frequency, and range data. This allows an aircrew to localize a submarine and build a track much more quickly.

Fast and accurate TMA comes at a price, however. Active sonar is very overt. A submarine can detect active sonobuoy emissions at a range much further than the sonobuoy will detect the submarine. This allows a submarine crew to plot the location of the active sensor and gives them time to evade the sonobuoy. Submariners have a variety of ways to degrade active sonar returns and evade prosecution with active sonobuoys.

When quantifying active sonar performance, we find that detection range is limited by one of two modes. In the first scenario, detection range is driven by AN levels in the water column. This is known as an Ambient Noise Limited case. In the second scenario, detection range is limited by a phenomenon known as Reverberation. In the Reverberation Limited case, sound energy scatters off of objects in the water column, including waves, the ocean floor, and marine life. The reflections off these objects tend to drown out the return echo from the submarine.

Many of the factors that quantify active sonar performance are already familiar to us from our discussion of passive sonar. Let's examine the Ambient Noise Limited active sonar equation to better understand the differences:

$$SE = SL + TS - 2TL - NL + DI - RD \text{ (Ambient Noise Limited)}$$

We are already familiar with the values SE, RD, NL, and DI. In this equation, SL is the strength of the sound energy emitted by the active sonobuoy, not the sound energy generated by the submarine. Unlike the passive sonar equation, the active sonar equation doubles the value TL. This is because sound must travel from the sensor to the target and back to the receiver. This doubles the amount of energy stripped away through TL.

The value TS is an abbreviation for Target Strength. This is a measure of how much sound energy will be reflected by the submarine back towards the target. TS is a function of the size, shape, and materials used to build the submarine. TS is aspect dependent, meaning that it varies with the direction of arrival of the incoming sound energy. Typically, TS is strongest at the beam aspect of the submarine and tapers off at the bow and stern. In theory, this leads to a characteristic "butterfly shape" plot of TS, with high values off the beam, small values at the bow and stern, and slightly larger values at the quarters.

This idealized TS configuration is not always observed in testing. Actual in-water trials have shown there is often great variation in observed TS values, not corresponding with values inferred during modeling.[8] The strength of a sonar pulse return is dependent on complicated interactions, as sound reflects and is at times reinforced or canceled out by the arrangement of the boat itself. It is believed that the pronounced increase observed at the quarters of "butterfly shape" TS plots are due to complicated interaction of sound reflecting off the fuel tank and bracing structure located outside of the pressure hull of post-World War II diesel submarines. These boats were the targets used during research to compile the available open source literature on submarine TS. Such behavior is not likely to occur in modern diesel and nuclear submarines where the fuel and ballast tanks are inside the pressure hull. Regardless, the high level of security surrounding submarine design means there is very little, if any, open source literature describing TS values of modern submarines.[9]

As a result of TS being aspect dependent, the strength of active sonar returns are not only due to the environment and range from the sensor to the target but also the orientation of the submarine to the sensor. TS is also dependent on the frequency of the active sensor being used. Submarine hulls are frequently coated in reflective synthetic material known as "anechoic tiling." This tiling is meant to absorb active sonar, reducing the TS against a particular frequency or set of frequencies commonly used by ASW forces.[10] As a result, TS should be expressed for the particular frequency and waveform used by the ASW aircraft's sensors, not just as a general value.

To understand the Reverberation Limited active sonar case, we need to quantify a value called Reverberation Level (RL). This is a measure analogous to NL, which seeks to quantify the amount of reverberation caused by sound energy reflecting off of waves, the ocean bottom, marine life, and particulate matter near the sensor.

$$SE = SL + TS - 2TL - RL - RD \text{ (Reverberation Limited)}$$

Just as in the case of passive sonar, aviators can use predicted environmental measurements and generate TL curves. By substituting these values into either of the active sonar equations, they can predict SE as a function of range and determine how far from a sensor they can detect a submarine.

Active sonars use a variety of waveforms and frequencies to detect their targets. Just as in the case of passive systems, the wavelength generated by active sonars is tied to the physical size of the array. As a result, air-dropped sonobuoys are unable to generate low-frequency sound energy because their arrays are too small. Lower frequencies generally have longer ranges and are less attenuated by water. Higher-frequency emissions generally produce more accurate location information but lose energy far more quickly, leading to shorter detection ranges. As mentioned above, different frequencies will be more or less effective against anechoic tiling depending on the material composition of the tiles.

Active sonar pulses are shifted in frequency by the relative motion of the submarine as they reflect off the hull and back to the receiver. This allows aviators to infer relative motion and conduct TMA. A submarine moving towards an active sonar pulse will cause the returned signal to shift up in frequency due to the Doppler effect. Doppler is a key tool for acoustic operators to help discriminate between returns caused by the ocean bottom and a submarine. Some active sonobuoys use frequency-modulated (FM) pulses to artificially generate Doppler and help detect low-speed submarines or boats hovering dead in the water.

Active sonar will generate returns off various parts of the submarine hull as well as the wake.[11] Active sonar systems have minimum as well as maximum detection ranges. Sonobuoys calculate distance by measuring the time from when a sound pulse left the array to when it returned. If the echo returns to the sonobuoy before the array has ceased emitting, the receiver will not be able to detect any returning sound energy. This sets a minimum detection distance. Aviators refer to this phenomenon as "range gate blanking." They may use very short-duration sound pulses to minimize the size of the range gate.

Submariners are well aware of the capabilities of active sonobuoys. They can detect sonobuoy emissions and plot their general location. A submarine can reduce its detection range and perhaps disappear from a sensor's field of view by simply turning its bow or stern towards the sensor to present the lowest TS value. It can also degrade the sonobuoy by operating at low speed, hovering, or resting on the ocean bottom to present minimum Doppler and hide in the clutter of the ocean floor.[12]

Multistatic Active Sonar

Up to this point, we have restricted our discussion of active sonar to cases in which the transmitter and receiver are co-located. This arrangement is called a monostatic active sonar. ASW aircraft also utilize active sonar systems where transmitters and receivers are located in different locations. Active sonars that transmit from one sonobuoy and receive an echo at another sonobuoy are known as bi-static active sonars. Sonars that emit from one sonobuoy and receive echoes at multiple receivers, possibly including the source, are known as multistatic active sonars.

Multistatic active sonar systems offer superior performance against quiet submarines in harsh water columns compared to monostatic systems. They present dilemmas for evading target submarines, offer improved detection geometries, and are more economical than monostatic systems.

Because of their long detection ranges, multistatic sonobuoy fields can be spread over large distances and offer good geometric opportunities to observe a target from multiple angles. Multistatic source sonobuoys are more expensive than receiver sonobuoys. This allows an aircraft to seed a multistatic sonobuoy field with similar numbers of sources but a larger number of inexpensive receivers. In doing so, aviators can generate large increases in field performance for moderate increases in costs, leading to an economic advantage. Last, submarines can easily detect monostatic active sonars and evade them by presenting low TS aspects and low Doppler. However, multistatic active sonobuoy fields allow the aircraft to ping from many different directions, presenting a geometric challenge that is difficult for a submarine crew to evade.

Multistatic systems also have their drawbacks. They require powerful and expensive source sonobuoys. They require more advanced signal processing compared with monostatic systems. They are vulnerable to high false alarm rates and require echoes be registered on multiple receivers to help drive down localization errors.[13]

While discussing monostatic active sonars, we referenced the relationship between target aspect, TS, and the strength of sonar echoes. With multistatic systems, the geometry is a bit different. When modeling multistatic active systems, engineers are concerned with a measurement known as the "bistatic angle." This is the angle between an arriving sound ray that hits a submarine and the reflected echo that travels to a receiver. When the bistatic angle is equal to 90 degrees, a strong echo known as a "specular return" is registered on the receiver.[14] Multistatic fields are spread over large areas, raising the chances that a single or many specular returns will be observed. These strong returns help aircrews refine the position of the target submarine as they localize.

As multistatic sonars emit acoustic energy, sound waves bounce off the target submarine, objects in the ocean, the ocean floor, and objects on the bottom such as shipwrecks. A large amount of energy returns to the receivers, resulting in many false alarms. Discriminating between the target submarine and spurious returns is a key part of effective multistatic sonar employment.

To discriminate between the actual target and false returns, operators observe the signature and aural characteristics of the echo. A return from a submarine looks and sounds different than an echo produced by a shipwreck or seamount. This allows aviators to differentiate between high-confidence and low-confidence returns. Multistatic systems also benefit from human-machine teaming, where algorithms sift through returns and filter out low-confidence returns in order to reduce false alarm rates.[15]

High false alarm rates have distinct tactical implications for multistatic sonar systems. Multistatic emission plans are constructed and modeled under the assumption that a certain amount of sound energy will be injected into the ocean, yielding a chance of a certain number of detection events. If aircrews are breaking off the emission plan to localize on returns that prove to be false alarms, they are lowering the chance that the emission plan is completed. If the emission plan is not carried out to the end, the cumulative detection probability (CDP) of the search pattern will suffer. This incentivizes engineers to build multistatic sensors and acoustic processors that limit false alarm rates. ASW mission planners have to build enough time into search operations to allow aircrews to localize on suitable high-confidence returns, convert the datum into a high-quality track or realize the return was false, and return to searching and complete the emission plan.

Just as monostatic active sonar systems suffer from range gate limitations, so too do multistatic systems. Multistatic sources emit a pulse of sound energy for a certain duration. If the first parts of the wave echo off the target before the sensor stops transmitting, any echo will be hidden by the source itself. This phenomenon is known as "direct blast" and the resulting area where the target may be hidden is called the "blanking zone."

Early multistatic sonar systems used explosive sound sources to produce echoes off the target submarine. These sound sources were very powerful, but the energy emitted was not very precise. More recent multistatic systems have utilized hydrophones that can produce coherent pulses of sound waves. They can tailor the waveform of the sonar pulse to the tactical situation at hand. For example, a continuous wave emission is sensitive to Doppler and is therefore well suited to a scenario when the target is likely to be moving towards or away from the source and receiver pair. Conversely, a FM waveform is best when the target is moving parallel to a source and receiver pair.[16]

Like monostatic active sonars, multistatic systems are not capable of classifying and identifying targets. They are best used for developing a small search zone known as an Area of Probability (AOP) that can then be seeded with monostatic active or passive sonobuoys to develop a high-quality track while also identifying the target.

Early multistatic sonar systems were developed in an era where computing power was much more limited. As a result, these systems focused on employing symmetrical sonobuoy patterns with simple geometry. This made modeling the performance of the sonobuoy field far easier and less computationally intensive. From a theoretical standpoint, the optimal multistatic field would take advantage of non-uniformity in the environment. More sources would be placed in regions of the operating area where sound propagation is degraded while fewer sources would be placed in areas where sound propagates more easily.[17] Sources and receivers would also be positioned in response to the predicted path of the target and its expected tactics.

Recent advances in computing power have made it realistic to model and select arrangements of sources and receivers that are tailored to the water column and tactical scenario at hand. Positions for sources and receivers and the types of waveforms used are chosen based on what yields the best detection probability during search and the most accurate location data during tracking.[18]

Magnetic Anomaly Detection

Submarines are large objects built mostly of ferrous metal, which tends to produce a magnetic field or become magnetized during operations. The magnetic field of the submarine is substantial enough that a low-flying aircraft carrying equipment to measure magnetism will detect a large and rapid variation in the local magnetic field as it passes overhead.

Unlike radiofrequency or light energy, magnetism is a field rather than a traveling wave phenomenon. This means that it is impossible to build an antenna to focus magnetic fields and reject background magnetic noise. The intensity of a magnetic field decreases proportionally to the cube of the distance. As a result, MAD is limited to short-range applications and airborne MAD ranges are less than a few thousand feet.[19]

The short-range nature of MAD limits its utility as a search sensor. Modern ASW aircraft generally use MAD equipment during localization and to corroborate passive sonar contact when dropping a torpedo. Range limits force ASW aircraft to fly as low as possible when using MAD equipment in order to detect a submarine over a wider swathe of ocean and down to a deeper depth.

The earth's magnetic field is complex, with numerous local variations caused by ore deposits and shipwrecks. This effect is more pronounced in shallow water because the magnetic deposits in the earth's crust are physically closer to the detector. Magnetic surveys can identify areas of magnetic deposits and display their intensity using contour charts similar to those used to depict elevation.[20]

The planet's magnetic field is also affected by solar storms, which bombard the atmosphere with charged particles. These storms vary in frequency, with peaks occurring during high points in the sun's 12-year cycle. During periods of low solar activity, magnetic storms disrupt MAD equipment about 3 percent of the time. During periods of high sun activity, solar storms occur about 15 percent of the time. This disruption phenomenon is much more pronounced at high latitudes than it is near the equator.[21]

An aircraft flying over a submarine will register a characteristic signature on MAD equipment. The shape of the signal is driven by the motion of the aircraft and that of the submarine. The speed of the aircraft will vary the shape of the signal, with higher speeds producing more pronounced "spikes" on MAD equipment. The course of the aircraft relative to that of the submarine will also affect the shape of the MAD signature. Lastly, the heading of the submarine in relation to the earth's magnetic field will also drive how the anomaly appears on the aircraft's MAD equipment.[22]

Solar storms and increased magnetic noise do not mimic submarine-generated MAD signals, but rather tend to mask the actual signal in background noise. MAD is also influenced by psychological factors. An operator who anticipates seeing a MAD signal while flying over an AOP will often mistake naturally-occurring variations for the submarine anomaly. In high-pressure situations, aviators often "see what they expect to see."

MAD can also be degraded by the ASW aircraft itself and the equipment used by the crew. Radio transmissions tend to disrupt MAD equipment. The magnetic structure of the aircraft and its engines also influence the MAD sensor. As a result, most ASW aircraft carry MAD sensors in tail booms, as far from most of the fuselage structure as possible. Maintenance personnel carefully calibrate the MAD sensors to minimize the effect caused by the ferrous metal in the aircraft.

LIDAR

The ocean is very effective at absorbing light, with the exception of a small portion of light waves in the blue-green portion of the visual spectrum. ASW forces have experimented with blue-green lasers to transmit pulses of energy into the water and measure the increased or decreased amount of return energy caused by light reflecting off a submarine hull or being absorbed by the dark hull. These systems are known as Light Detection and Ranging sensors, abbreviated as LIDAR.

The main limiting factor that drives LIDAR detection effectiveness is the depth of the submarine. Light energy, even in tightly focused, coherent laser pulses, is scattered and attenuated by water. Below 150 feet, there is simply not enough light returned to register the presence of a submarine.

Attenuation is strongest in areas where large amounts of particulate matter are present. The first 300 feet below the surface attenuate light more strongly than water at greater depths. The open ocean tends to have less particulate matter and therefore is more transparent. Low turbidity is the best predictor of whether LIDAR will be effective in a particular water column.

When a laser pulse is fired at the ocean surface, part of the light reflects off the surface and part of the light penetrates the water. The challenge for engineers building a LIDAR system is discriminating between the high-intensity return caused by reflections off the ocean and the low-intensity returns caused by the ocean bottom or the submarine.[23]

Other environmental factors aside from turbidity affect LIDAR performance as well. High winds cause whitecaps, which in turn cause false returns. Low winds result in a glassy ocean, and there may not be sufficient surface ripples to reflect enough light to allow the processing algorithm in the LIDAR to discriminate between returned energy from the surface and returned energy from objects below the surface. Sun glint can also blind LIDAR systems. As a result, LIDAR employment for bathymetric surveys or ASW is generally conducted during the mornings or afternoon, rather than around noon, when the sun is directly overhead.[24]

Despite the limited depth performance of LIDAR, it is a proven system. It has been used extensively for mapping bathymetric features in shallow water and for detecting marine life. It is likely that the Swedish Navy has operationally employed LIDAR systems since the 1980s to search for Russian submarines in the shallow Baltic Sea.[25] LIDAR systems have also been deployed to detect mines and underwater objects.

Laser and power-generation systems are capable of producing a suitably powerful signal from either aircraft or satellites. The largest obstacle to employing orbital LIDAR is that clouds and fog rapidly attenuate light energy. Large portions of the ocean are covered by clouds as much as 60 percent of the time. For certain northern latitudes of particular interest to ASW operations, like the GIUK Gap, clouds may be present up to 80 percent of the time. As a result, LIDAR systems are deployed on aircraft, which can fly below cloud cover.[26]

Submarine Wave and Wake Detection

As a submarine moves through the ocean, it disturbs the water around it in ways that may reveal its presence. Submarine hulls produce surface waves, underwater wakes, and internal waves that may be observed. A boat may also interact with the water in such a manner that differences in surface temperature become apparent on IR sensors.

A submarine moving underwater will displace water around it, causing a near field surface wave. The near field wave is often referred to as a "Bernoulli hump" because the displaced water will cause a slight rise in the ocean surface. The height of the near field wave is most strongly governed by the depth and speed of the submarine. A shallow, fast-moving submarine will generate a larger Bernoulli hump compared with that of a slow, deep boat.

Unfortunately for ASW forces, the Bernoulli hump is small compared to wave action, and is easily swallowed up by wind-driven waves. An *Ohio* class SSBN moving at 20 knots and a depth of 100 feet will generate a near field wave about 6 inches high. At more realistic patrol speeds of five knots and 300 feet, the Bernoulli hump will be less than 1/25th of an inch high. ASW forces cannot realistically detect a disturbance that is at most several thousand feet long and is three orders of magnitude smaller than surface waves.[27]

Moving submarines also generate far field surface waves, which spread out in a manner similar to the wake of a surface ship. The divergent and transverse waves are bounded by an angle of 19.5 degrees.[28] While the far field waves persist over much longer distances than the Bernoulli hump, they are much smaller in magnitude. Like the near field, the magnitude of far field waves is strongly influenced by submarine speed and depth. A boat running at 12 knots at a very shallow depth will produce far field waves only 1/10th of an inch high. At typical patrol speeds and depths, the surface waves are not apparent at all.[29]

A submarine will also produce an underwater wake, due to the movement of its hull and the motion of its propeller. As the boat moves forward, the turbulent area behind the submarine will grow vertically and horizontally, eventually reaching a maximum vertical extent of 1.3–3 times the diameter of the hull. As warm and cold water are entrained into the wake, the wake will continue to spread out horizontally.[30]

The turbulence of the wake will tend to decay as time goes by until the disturbances are no longer measurable. Acoustic tests indicate that wakes can be detected at least 1,500 feet behind the boat. Scale model testing has demonstrated that wake turbulence can be detected several miles downstream of the boat before the disturbances completely die away.[31]

Wake detection was an area of great interest to the Soviet submarine fleet.[32] Unable to match the quality of Western sonar arrays and signal processing, the Soviets sought to leverage non-acoustic detection mechanisms.[33] Several classes of Soviet SSNs went to sea equipped with wake detectors to register the turbulence caused by a foreign submarine.[34]

Wake detection is becoming an area of interest again in the modern era due to advances in autonomous vehicles and increasingly quiet submarines.[35] Long-duration UUVs are ideal platforms to deploy into a patrol area and detect the wakes of passing submarines. These detection events serve more as cueing information than they do tracking opportunities due to the limited kinematic capabilities of UUVs.

Bioluminescence

A submarine moving through the water will disturb marine life as it does so. As the boat's hull or turbulent wake disturbs or kills certain species of zooplankton and dinoflagellates, these organisms will emit light. This phenomenon is known as bioluminescence. In theory, the glow caused by these tiny animals can be detected by an airborne or space-based sensor.

The intensity of light caused by bioluminescence is driven by a series of factors, including plankton population density in a particular location, the types of species disturbed by the submarine, the season and time of day, and the speed of the submarine. The types of plankton that cause bioluminescence tend to cluster in the upper parts of the water column, with maximum bioluminescence values occurring at depths from 150 to 450 feet deep. These species are much more prevalent in the shallow waters on the continental shelf and are much more numerous between the equator and 60 degrees of latitude than they are in the polar oceans. Plankton tends to emit more light at night and less light during the day.

The ability to observe bioluminescent light is driven by three primary factors: cloud coverage, depth, and background light levels. Sea fog and clouds are very prevalent over most of the ocean, with up to 60 percent of the ocean being covered by one of the two at any given time. This makes orbital observation of bioluminescence a non-starter but does not rule out observation by an ASW aircraft flying below the clouds.

Light caused by marine life must travel through the water to the surface and then through the atmosphere to a sensor, losing intensity as it does. As a result, depth is the primary driver that governs how strong a bioluminescent signal will be when it reaches the surface.

Once light has reached the surface, a sensor must discriminate it from background light caused by the sun or moon. The intensity of light varies greatly from its maximum intensity at noon to minimum intensity at night with a black sky and low moon illumination. Bioluminescent light will not be observable during the day. At night, it may observable if the boat is operating at shallow depth.

By analyzing typical plankton densities and behavior, it becomes clear that even in a best-case scenario, a boat below 150 feet will generally not be detected, even on a dark night under ideal circumstances. Disturbed marine life may be carried close to the surface by a submarine wake, but these wakes are very unlikely to reach the surface if the boat is deeper than 150 feet. As a result, submarines can avoid being detected due to bioluminescence simply by operating at moderate depths. The relatively limited field of view of EO and IR sensors onboard ASW aircraft does not help the aviators.

Last, it must be mentioned that feeding fish and surface ships also cause bioluminescence. Determining what signals are bioluminescence caused by a submarine and what signals are caused other factors is a very challenging task. The above reasons combine to make detection of bioluminescence a very limited method of ASW search.

Notes

1. Flooding an area with radar while operating a HVU or convoy at high speed have been traditional tactics against quiet conventional submarines for decades. High speed forces a submarine to drain its batteries to approach a HVU during an attack. Radar flooding makes submarines reluctant to snorkel and easily detectable when they do expose a mast. Owen Cote and Harvey Sapolsky, "Antisubmarine Warfare After the Cold War" (Cambridge, MA: Massachusetts Institute of Technology Security Studies Program, June 1997), http://web.mit.edu/ssp/publications/conf_series/ASW/ASW_Report.html. (Accessed 27 August 2016).

2. While passive sonar itself is covert, sonobuoys need to transmit the information they receive to ASW aircraft flying above. This is done using a radio transmitter in each sonobuoy and a receiver antenna in the aircraft. In theory, a radiofrequency emission by a sonobuoy can be detected by a submarine or intelligence gathering platform nearby. Sonobuoys also make noise as they enter the water, raising the chance that a submarine with a sensitive passive sonar can detect sonobuoy drops nearby and infer the presence of an ASW aircraft.

3. Radiated Noise Ranging for Submarines (Perth, Western Australia: Xenthos, undated), p. 6, http://xenthos.com.au/Ranging%20Paper.pdf (Accessed 21 May 2016).

4. Ibid.

5. Ibid.

6. "Reference Publication (RP) 33: Fleet Oceanographic and Acoustic Reference Manual" (Stennis Space Center, MS: U.S. Navy Naval Oceanographic Office, April 1999), p. 29.

7. Roger A. Holler, "Drag Measurements of Long Cables in the Ocean" (Warminster, PA: Naval Air Development Center, 26 March 1984), p. 4.

8. Robert J. Urick, *Principles of Underwater Sound* (Westport, CT: Peninsula Publishing, 1983), p. 311.

9. Ibid., pp. 318–19.

10. Ibid., p. 321.

11. Ibid., p. 310.

12. These concepts can all be deduced from first principles. Detection range is a function of TS, so any efforts to reduce TS will naturally reduce detection range in a particular scenario. In a similar manner, relative motion and Doppler are key tools used by active sonar systems. Any action that presents low Doppler to an active sonar will make it more challenging to discriminate between returns from a submarine and background clutter.

13. Alan R. Washburn, "A Multistatic Sonobuoy Theory" (Monterey, CA: Naval Postgraduate School, 2010), p. 10.

14. When the bi-static angle is 90 degrees, a large rise in TS occurs, causing the strength of a specular return to grow an order of magnitude or more compared to returns with less advantageous geometry. Because the geometric relationship of the source and receivers are known, the heading of the submarine can be inferred, although two ambiguous solutions will be presented. As tracking continues, the correct heading can be deduced from bearing drift. Doug J. Grimmett, "Automatic Identification of Specular Detections in Multistatic Sonar Systems," Proceedings of the IEEE Oceans 2009 Conference (Biloxi, MS: October 2009).

15. Doug J. Grimmett, "Reduction of False Alarm Rate in Distributed Multistatic Sonar Systems Through Detection Cueing," Proceedings of the IEEE Oceans 2007 Conference (Aberdeen, Scotland: June 2007).

16. O. Erdinc, P. Willett, and S. Coraluppi, "Multistatic Sensor Placement: A Tracking Approach" (Storrs, CT: University of Connecticut, 4 December 2007), p. 6.

17. Donald R. DelBalzo and Kevin Stangl, "Design and Performance of Irregular Sonobuoy Patterns in Complicated Environments," Proceedings of the IEEE Oceans 2009 Conference (Biloxi, MS: October 2009).

18. For ease of modeling and employment, most operational multistatic active sonobuoys are deployed in "posts" where a source is paired with a receiver. From a purely mathematical perspective, there are many ways to improve system performance using various arrangements of sources and receivers. These relationships are driven in part by the cost relationship between sources and receivers. The benefit of employing complicated geometric sonobuoy arrangements must be balanced against the penalty of fuel and time required for the ASW aircraft to lay the sonobuoy field. For more details on multistatic field geometry, see S. Ozols and M. P. Fewell, "On the Design of Multistatic Sonobuoy Fields for Area Search" (Edinburgh, South Australia: Defense Science and Technology Organization, June 2011).

19. Tom Stefanick, Strategic Antisubmarine Warfare and Naval Strategy (Lexington, MA: Lexington Books, 1987), pp. 184–8.

20. James A. Brennan and Thomas M. Davis, "The Influence of the Natural Environment on MAD Operations" (Stennis Space Center, MS: U.S. Navy Naval Oceanographic Office, September 1969), pp. 20–2.

21. Stefanick, Strategic Antisubmarine Warfare, p. 186.

22. Brennan and Davis, "The Influence of the Natural Environment on MAD Operations," pp. 6–20.

23. As a method of comparison, it's worth considering that the magnitude of the energy reflected off the surface may be six orders of magnitude greater than that reflected off an object underwater. G. C. Guenther, G. A. Cunningham, P. E. LaRocque, and D. J. Reid, "Meeting the Accuracy Challenge in Airborne LIDAR Bathymetry," Proceedings of the EARSek.SIG Workshop, 16–17 June 2000.

24. John Olav Birkeland, "The Potential for LIDAR as an Antisubmarine Warfare Sensor" (Glasgow, Scotland: University of Glasgow, 2009), pp. 49–50.

25. "Interest in Laser Radar Grows After Sub Affair," Svenska Dagbladet (17 November 1981).

26. Stefanick, Strategic Antisubmarine Warfare, p. 207.

27. Ibid., pp. 193–4.

28. Ibid., p. 195.

29. Ibid., p. 197.

30. Ibid,, pp. 197–9.

31. I. Pelech, G. G. Zipfel and R. L. Holford, "A Wake Scattering Experiment in Thermally Stratified Water," Journal of the Acoustic Society of America, Vol. 73, No. 2 (February 1983), pp. 528–38.

32. David Hambling, "How the Soviet Union Snooped Waters for Enemy Subs – Without Sonar," Popular Mechanics (23 October 2017), https://www.popularmechanics.com/military/navy-ships/a28724/submarine-sonar-soks/ (accessed 24 December 2017).

33. "SR IR 72-13-S: Soviet Antisubmarine Warfare: Current Capabilities and Priorities" (Washington, D.C.: Central Intelligence Agency, September, 1972), pp. 63–81; Prospects for Soviet Success in Improving Detection of Submarines in the Open Ocean" (Washington, D.C.: Defense Intelligence Agency, November 1974), pp. 3–7.

34. Certain Russian Project 671RTM/VICTOR III class SSNs and Project 971/AKULA class SSNs are equipped with non-acoustic sensors systems that appear intended for wake detection. These systems have been referred to as Kolos and SOKS in open-source literature. David Ross, The World's Greatest Submarines: An Illustrated History (London: Amber Books, 2017), p. 138.

35. Since 2017, British SSNs have been occasionally sighted with what appear to be wake detection sensors on their sails or bow. HMS *Trafalgar* (S107) was photographed in 2018 with items resembling non-acoustic sensor inlets on her sail. HMS *Talent* (S92) was similarly photographed in May 2019 with probes resembling wake detection sensors on her bow. Joseph Trevithick, "Royal Navy Sub Appears in Gibraltar Equipped with a Wake Detection System," *The War Zone* (8 May 2019), https://www.thedrive.com/the-war-zone/27913/royal-navy-sub-appears-in-gibraltar-equipped-with-a-wake-detection-system.

Chapter 8

Tactical Oceanography and Environmental Analysis

To be effective at planning and searching for a submarine, aviators must understand the environment in which they are operating. In this chapter, we will discuss the basics of tactical oceanography and environmental analysis. Since modern ASW aircraft rely overwhelmingly on passive and active sonar to find a target submarine, we will focus the majority of our discussion on acoustics. We will discuss how sound propagates in the water and how the properties of the water column affect its movement.

Tactical oceanography is an exceedingly complex subject. There is an enormous amount of literature covering sound propagation, AN in the ocean, TL, and other topics. This chapter is intended to serve merely as an introduction to environmental analysis. Readers who are interested in more detailed discussions should make use of the works listed in the notes to learn more.

When searching for a submarine, aviators try to answer the question "where should I put a sensor to give the best chance of finding the target?" They like to know how far from a sensor a submarine could be heard. They also find it useful to know what sort of effect variations in parameters such as AN will have on their ability to detect their target. By the end of this chapter, the reader will be armed with a basic understanding of how changes in parameters such as temperature, depth, salinity, and humidity influence the propagation of acoustic and radiofrequency energy.

Transmission Loss

A crew of aviators searching for a submarine are intensely interested in how far from a sonobuoy they can detect their target. One of the largest factors impacting detection range is the rate at which sound energy is lost as it moves through the water from the submarine to the sensor. From our discussion of sensor theory, we know this parameter as Transmission Loss (TL). TL is the result of two phenomena: losses due to spreading and losses due to attenuation.

Sound is wave motion produced by a vibrating object. In the ocean, these sound waves are transported through the water column using seawater as the transmission medium. As sound energy spreads out from the submarine, it covers an increasingly large geometric area. As the sound waves move further and further from the sound source, the intensity of acoustic energy at any one point becomes smaller. This is known as "spreading loss." We can break spreading loss down further into one of three different spreading mechanisms: spherical spreading, cylindrical spreading, and dipolar spreading.

Spherical spreading is the simplest form of spreading loss. Imagine a sound source in a homogenous medium. As acoustic energy spreads out from the source, it forms a spherical shell. The energy is spread over an increasingly large surface, whose area we can describe as $A = Pi*r^2$ where r is the range from the sound source. When spherical spreading dominates, we find that the spreading loss (L_S) from the source to the receiver can be calculated using the mathematical expression $L_S = 20\log(r)$. As a result, a doubling of range from a target to a sensor will cause a 6 dB loss of intensity due to spherical spreading.

Now, let us imagine a situation where a water column has an upper and lower bound. This means that acoustic energy spreads out radially from a source but is capped on the top and the bottom by barriers such as the seafloor and the ocean surface. This type of spreading loss is known as cylindrical spreading. The acoustic energy spreads over an area equivalent to $A = 2*Pi*r*h$, where r is the range from the source and h is the height between the upper and lower bounds. We find that spreading loss from cylindrical spreading is equivalent to $L_S = 10\log(r)$. As a result, a doubling of range from a target to a sensor will lead to a 3 dB loss of sound intensity due to cylindrical spreading. Because cylindrical spreading confines acoustic energy, much less acoustic energy will be lost per unit of increased range when compared with spherical spreading. Cylindrical spreading leads to tactically exploitable extended-range sound propagation paths which we will discuss later in this chapter.

The third type of spreading is known as dipolar spreading. At certain points in the water column, sound speed is higher than that found in surrounding areas. Sound waves always seek to enter a lower-speed portion of a medium, causing them to bend away from areas of higher speed. A layer of water that has a high sound speed compared to the water above and below will cause sound waves to bend above and below. This phenomenon is known as dipolar spreading and can be found near a region of the water column known as the "layer" or "thermocline." Under dipolar spreading, the spreading loss is equivalent to $L_S = 40\log(r)$. Therefore, a doubling in range results in a 12 dB loss of intensity. This very high rate of energy dissipation is advantageous to submarines and disadvantageous to ASW aircraft. Submarines can take advantage of high spreading loss rates by operating near areas of maximum sound speed, causing their radiated noise to be dissipated more quickly and helping them evade ASW forces.[1]

Attenuation

Now that we are familiar with spreading losses, we may turn our attention to the other source of TL, attenuation. Attenuation is the sum of losses due to three effects: absorption, scattering, and leakage of acoustic energy out of sound channels.[2] Absorption is the process of converting acoustic energy into heat, which leads to losses in the intensity of the sound wave. Scattering refers to wave energy being reflected away from the original direction of travel of the sound wave, leading to less acoustic energy arriving at a sensor. The structure of the ocean forms certain sound channels, where acoustic energy is ducted, allowing it to enjoy lower TL rates and to travel for long distances. A certain amount of energy leaks out of these sound ducts, leading to the final attenuation phenomena, leakage out of the duct.

As a sound wave moves through the seawater medium, it creates compressions and rarefactions in the water. This process, in turn, causes some of the wave's energy to be converted into heat in a process known as absorption. Absorption is caused by viscous effects and the ionic relaxation of particulate matter dissolved in seawater. As sound waves compress and expand seawater, they are opposed by the natural resistance of water to be deformed. When water molecules are disturbed, they need time to "flow" back into their lattice crystal structure. The movement of the molecules back into the lattice generates heat.[3] Seawater also contains a variety of dissolved ionic salts. As sound waves move through the water, they disturb magnesium sulfate ions, which dissociate and cause further absorption losses.[4]

Rates of absorption are correlated with frequency, with higher frequencies showing higher rates of absorption loss. Absorption is inversely correlated with depth, with measurements in deep water showing lower losses due to absorption.[5] This lower rate of attenuation in the deep ocean contributes to the formation of deep sound propagation paths which we will discuss later.

As sound waves move outwards, they come into contact with discontinuities that redirect acoustic energy in directions other than the path of the sound wave. Energy reflecting away from the original direction of travel results in a less intense sound wave being observed by a sensor. This type of attenuation is known as scattering. Acousticians also use the term "reverberation" to describe the process of sound waves being scattered.[6]

Scattering is caused by sound waves interacting with the ocean surface, the seafloor, and matter suspended in the water column. Surface reverberation is the result of sound waves interacting with the air-water interface, which itself is disturbed by wave action and bubbles. Sound waves that hit the seafloor are absorbed by the bottom and also reflected in various directions. Finally, sound waves reflect off marine life and inorganic matter in the ocean itself, causing a phenomenon known as volume reverberation.

Volume reverberation seems to be caused primarily by scattering sources that are biological in nature. They are not evenly distributed in the water column. Rather, these groupings of marine life tend to congregate in one or several layers known as the Deep Scattering Layer (DSL). The depth and height of the DSL, its overall scattering strength, and its scattering strength with respect to frequency vary greatly depending upon the location, season of the year, and time of day.[7]

At frequencies from 2 to 10 kHz, the DSL scattering strength appears to be primarily due to acoustic energy reflecting off of the swim bladders of fish. Above 20 kHz, the DSL scattering appears to be primarily due to reflections off of zooplankton and the small organisms that feed on them. The DSL typically displays a strong diurnal variation, meaning that it tends to migrate to a shallower depth at night and move to deeper water during the day. The DSL depth shows a rapid change at sunset and sunrise, as marine life migrates upwards near sunset and returns deeper in the water as sunrise approaches.[8] Volume reverberation typically decreases at truly great depths, as there is generally much less life in the deep ocean because of lower levels of nutrients and lack of sunlight.[9]

Sound Refraction

In a simple world, sound would move through the ocean from a submarine to a sensor in a straight line. Unfortunately for aviators, sound rarely, if ever, moves in a straight line in the ocean. However, several simple physics concepts can help aviators and planners understand and predict how sound moves in the ocean.

A concept known as Snell's Law is used to describe the propagation of wave phenomena in heterogeneous mediums, of which the ocean is one. Snell's Law governs the angle at which a wave arrives, known as the angle of incidence, and the angle with which it continues onwards in a new medium, known as the angle of refraction. Snell's Law states that the ratio of the sines of the angles of incidence and refraction is equivalent to the ratio of the velocity of the wave phenomenon in each medium.

More simply put, sound moves at different speeds in different parts of the ocean. Sound will always tend to bend towards and seek out areas of the ocean where the local speed of sound is lowest. Therefore, sound tends to bend, or be "refracted" in various non-linear paths. If the speed of sound at different locations in the ocean is known, aviators can model how the sound will refract and predict the non-linear pathways over which the sound waves will travel.

To understand the concept better, let us compare two water columns, one of which is idealized and the other being more realistic. Consider a sound source sitting in a part of the ocean where the properties of the seawater are such that the speed of sound is constant everywhere. This is known as an isovelocity condition. As sound waves radiate from a source, they move outwards in a spherical direction. The sound waves have no tendency to bend, but rather move out radially. Unfortunately, isovelocity conditions are exceedingly rare in the ocean.

Now let's consider a slightly more complicated water column. The speed of sound in this area is lowest at the surface and increases in a linear manner until it reaches a maximum value at the bottom. Because the speed of sound is slower near the surface compared to the bottom, sound that radiates outwards from a source will be bent upwards, or refracted, towards the area of slower sound speed near the surface. **Figure 3** shows sound propagation under isovelocity conditions and under a positive sound speed gradient caused by isothermal conditions.

To quantify and simulate how the sound waves refract, acousticians can model the water column as consisting of many separate layers, each having a particular sound speed. The acousticians can define a particular sound wave moving outwards from the sound source in an initial direction as a "ray." As the sound wave moves through each water layer, Snell's Law can be used to calculate the angle of incidence and the angle of refraction. A sequence of interactions with many layers can be assembled to produce a "ray path," or a graphic description of the path that a sound wave will take through the water due to the effect of refraction.

Typically, scientists and ASW aviators use computers to carry out ray tracing. From a mathematical standpoint, ray tracing is most accurate for high-frequency sounds with short wavelengths. However, from an operational standpoint, ray tracing is an excellent way for warfighters to visualize how sound is moving in the water column and where they should put their sensors to achieve the desired result.

Figure 4 shows a representative ray trace of an area of the ocean with a common SSP. We can see that the sound waves being produced by the submarine are clustered together into several distinct pathways. While we will discuss each particular pathway later in the chapter, it is important to understand that by simply measuring the SSP in the ocean, aviators or acousticians can use Snell's Law and ray tracing to generate a visual depiction of sound propagation paths. These ray traces are key tools to help aviators exploit the water column and choose the best tactic for a given situation.

Sound Wave Reflection

As sound waves come into contact with the ocean surface and the seafloor, they are reflected in new directions. The behavior of a wave as it is reflected is largely governed by the frequency of the wave (and therefore its wavelength) and the angle of arrival of the wave with respect to the surface or bottom.

The key parameter in modeling reflection off the ocean surface is how "rough" the surface is in relation to the angle of arrival of the sound wave and the size of the wave itself. A sound wave arriving at the surface at a small angle will tend to reflect cleanly off the surface, moving in the original direction of travel without much loss. A low-frequency wave with a large wavelength will also tend to be less affected by the roughness of the surface caused by wave action. For high-frequency sound waves with small wavelengths or for sound waves arriving at the surface at a large angle, the surface will act more as a scattering agent than as a reflector.[10]

As sound waves hit the seafloor, some of their energy is reflected and scattered back into the ocean while the remainder travels into the bottom and is lost. Much like the ocean surface, the tendency of the seafloor to reflect and scatter sound waves is highly dependent on the frequency of the sound wave and the angle of arrival. It is also highly dependent on the shape and composition of the bottom itself.

Acousticians use a measure known as acoustic impedance to describe the resistance of a material to conducting sound energy. If there is a sharp discontinuity between the acoustic impedance of water and the acoustic impedance of the ocean bottom, much of the sound energy will be reflected. This type of sharp discontinuity is found in areas with rocky bottoms. Conversely, an area where the seafloor more closely matches the impedance of the ocean will absorb more of the sound energy into the bottom.[11]

The relative roughness of the ocean bottom also impacts how sound waves are reflected. A smooth bottom will tend to reflect sound energy in the original direction of travel. A rough bottom, on the other hand, will tend to scatter incoming sound waves, directing some acoustic energy back towards the source rather than onwards towards the receiver.

Temperature, Salinity, and Pressure Effects on Sound Speed

Sound speed in the ocean is an expression of the temperature, salinity, and pressure at a particular point. Interestingly, scientists have not discovered any other parameters that seem to affect sound speed. With sufficient knowledge of these three values, an observer can calculate the sound speed. Calculating sound speed at various points and depths in the ocean allows ASW aviators to predict how sound waves will propagate from the target, through the ocean, and to their sensors. It also allows them to place their sensors in the best position to register a detection and helps them to quantify how likely they are to find their target.

Higher values of temperature, salinity, or pressure will each in turn increase the local sound speed. Of the three parameters, salinity tends to vary the least. Frequently, aviators will simplify their calculations of sound speed by making assumptions about the salinity level, holding this parameter constant, and only varying values of temperature and pressure.[12]

Sound speed itself is relatively difficult to measure and doing so usually requires costly and sophisticated instruments. Typically, aviators use environmental sampling buoys known as expendable bathythermographs (XBTs) to measure the ocean temperature at a variety of depths in a particular location. The bathythermograph buoy is dropped from an aircraft and lowers a probe through several thousand feet of seawater to record the temperature at periodic intervals. This information is transmitted back by radio to the aircraft overhead. Because pressure increases proportionally with depth, the pressure at a particular depth is always known. The aviators pick an appropriate salinity value for the area in question and use the temperature trace to infer the sound speed at each depth in the water column. This dataset is known as the sound speed profile, or SSP.[13] Accurate knowledge of the SSP is one of the most useful tools available to help an aircrew understand the ocean environment and the tactical situation.

The Deep Ocean Water Column

Let us examine a simple water column to better understand the relationship between temperature, pressure, and sound speed. First, consider a shallow part of the ocean. Mixing action by the wind and waves keeps the water well circulated from the surface to the seafloor. As a result, the temperature is uniform, a condition known as an isothermal water column. Assuming salinity is constant, the only variable that will affect sound speed is pressure. As we move deeper, the pressure will increase constantly. This leads to a situation where sound speed will be lowest at the surface and will increase linearly until it reaches a maximum value at the ocean bottom. This condition is common in shallow water and is shown in **Figure 5**.

Applying what we know about the effects of temperature, salinity, and pressure on sound speed, let us examine a typical temperature profile from the deep ocean and determine what the SSP will look like. To do so, we must familiarize ourselves with the typical vertical structure of the water column in the deep ocean.

Starting at the ocean's surface we find a layer of water that is constantly mixed by wind and wave action. This region is known as the mixed layer. The mixed layer is isothermal. Its depth is driven by the force of wave action. Layer depth tends to increase in the winter months, as stormy weather and larger waves agitate the water and drive surface water to a depth of 300 to 450 feet. During the summer months, the mixed layer tends to be shallower. At times, the mixed layer may be capped by a thin layer that is influenced by daily or even hourly changes in temperature and sound speed due to heating and cooling caused by the sun. This layer is sometimes referred to as the surface layer and is of much greater concern to surface ships than ASW aircraft.[14]

Below the mixed layer, temperature decreases as we travel deeper into the ocean. This area of temperature decrease is known as the thermocline. The lower portion of the thermocline grows cooler and cooler with depth until the water reaches a temperature of 39 degrees Fahrenheit (4 degrees Celsius). The lower regions of the thermocline are relatively stable in temperature, changing very little with the seasons. The upper regions of the thermocline shift in temperature, forming an area known as the seasonal thermocline. In the summer months, the seasonal thermocline heats up in response to heat transfer from the warmer and shallower mixed layer. During the winter, however, the seasonal thermocline tends to fuse with the deeper mixed layer.[15]

Below the thermocline lies an isothermal layer of cold water. This area of the ocean is a constant chilly 39 degrees Fahrenheit. It does not shift in temperature or respond to seasonal changes. Very little sea life is found in these deep regions, making them as barren as any desert on land.

A Deep-Water Sound Speed Profile
Armed with knowledge about the typical structure of the deep ocean, let us examine a representative deep-water SSP. For the sake of simplicity, we will

assume salinity is constant. In the mixed layer, temperature stays constant with an increase in depth. As we descend, the increased pressure will cause sound speed to increase as we move lower in the mixed layer. Somewhere at or near the bottom of the mixed layer we will reach a local sound speed maximum. This speed is known as the near surface sound speed maximum.

As we move lower into the thermocline, temperature will drop as we go deeper. Lower temperatures correspond with lower sound speeds, but we also know that increased depth means increased pressure, which will cause sound speed to increase. At first, the magnitude of the temperature decrease will dominate the effect of the increase in pressure due to depth. Eventually, however, we will reach the deep isothermal layer of cold water. Here, salinity and temperature are constant as we descend, but pressure continues to increase. This will result in a sound speed minimum somewhere near the interface between the thermocline and the isothermal layer. Below this sound speed minimum, sound speed will continue to increase until we reach the ocean bottom. **Figure 6** below shows a typical deep-water thermal profile and the corresponding SSP.

Shallow-Water Sound Speed Profiles

Let us examine a common shallow-water SSP. Aviators may frequently find that wind and wave action mixes shallow waters, leading to an isothermal water column. This causes an increase in sound speed with increasing depth, a so-called positive sound speed gradient. The positive gradient tends to bend sound rays upwards, until they reflect off the surface and are redirected downwards. This repeated refraction and reflection causes a ducting phenomenon known as a half channel.[16]

However, shallow-water SSPs are often far more complicated than half channel conditions or typical deep-water profiles. Shallow waters close to land often have highly variable salinity. This is usually due to run-off from rivers, where fresh water from the river mixes with salt water from the ocean to create zones of high salinity variations. These salinity variations in turn lead to large variations in sound speed and irregular SSPs. In short, shallow water conditions are often highly variable, change greatly over short distances, and may lead to unusual sound propagation paths.

Sound Channels and Ducting Phenomena

Now that we have discussed the structure of the water column, reviewed the effect of temperature, salinity, and pressure on sound speed, and understand reflection and refraction, we can examine the phenomena that cause ducting and lead to lower rates of TL. We know that variations in sound speed cause sound waves to refract into non-linear paths. In a similar manner, we understand that the ocean surface and seafloor will reflect and scatter sound in particular directions. Certain SSPs will cause refraction and reflection that in turn create pathways where sound waves are restricted to regions in the water column where acoustic energy enjoys lower levels of TL per unit distance. These pathways allow sound waves to travel from a submarine to a sensor over very long ranges.

These long-range propagation phenomena are known as ducts or sound channels. Each type of duct or sound channel plays a role in airborne ASW. Some enable surveillance sensors to detect submarines across entire ocean basins. Others improve the performance of short-range tactical sonobuoys. The four types of long-range propagation phenomena are: the surface duct, the half channel, the DSC, and internal sound channels. We will discuss each in detail in the following sections.

The Surface Duct

In cloudy, windy regions of the ocean the upper layer of the water column is mixed by wind and wave action. Cloud coverage prevents solar radiation from hitting the water and raising the surface temperature enough to generate a warm surface later. The wave action mixes the water, causing an isothermal layer, or mixed layer, to form from the surface down to a certain depth. The isothermal condition in the mixed layer causes a positive sound speed gradient to be formed near the surface. **Figure 7** shows an example of a temperature versus depth and SSP plot for a mixed layer condition.

Since the sound speed gradient is positive above the mixed layer depth, sound rays emitted by a sound source are bent upwards. Many of the rays are bent upwards till they hit the surface. These rays are reflected and scattered, with many of them reflecting downwards only to be bent upwards again by the positive gradient again.

Other rays radiating downwards from a sound source are bent upwards, but not enough that they stay above the bottom of the mixed layer. Once below the mixed layer, the rays encounter a negative sound speed gradient, which causes them to be bent downwards. Above a certain angle, however, all rays will be bent upwards and held above the bottom of the mixed layer. For a sound source located in the mixed layer, this phenomenon tends to constrain much of the sound waves in the mixed layer, which acts to form a sound channel. This sound channel is known as the surface duct.

The lowest ray that is bent back upwards and constrained in the surface duct is known as the limiting ray. The angle of this ray is dependent on the depth of the source and the strength of the sound speed gradient. Because sound energy below the limiting ray is bent downwards by the negative sound speed gradient below the layer, surface ducts tend to form an area known as a shadow zone. Very little acoustic energy propagates into the shadow zone, creating a situation where sensors in the shadow zone cannot detect the submarine operating nearby in the surface duct. **Figure 8** shows a ray trace of a surface duct, including the limiting ray and the shadow zone.

In reality, shadow zones are less pronounced than ray tracing makes them appear. Acoustic energy leaks out from the surface duct into the shadow zone. This is partially due to sound waves being scattered downwards rather than reflected at low angles into the surface duct.[17] The shadow zone will also be ensonified by rays moving downward that reflect off the seafloor and return upwards into the shadow zone.[18] Regardless, surface ducts tend to produce excellent long-range propagation and good detection performance for sonobuoys, assuming that the submarine and the sensor are both located inside the surface duct.

The depth of the mixed layer also has an important effect on what sound waves are trapped inside a surface duct. Sound frequency is intimately associated with sound propagation qualities. Low-frequency sound waves have large wavelengths and propagate well over long distances because they are less susceptible to attenuation. High-frequency sound waves are physically smaller and much more easily attenuated, leading to higher TL rates. However, AN tends to degrade low-frequency signals more strongly than high-frequency signals, because the ocean is generally noisier at lower frequencies and quieter at higher frequencies.

In order for acoustic energy to be constrained in a surface duct and travel over long distances, the wavelength of the sound wave must physically fit inside the duct. This leads to a phenomenon known as low frequency cutoff. Depending on the depth of the mixed layer, there will be a certain size of sound wave that is physically too large to fit in the duct. Above this frequency, sound will propagate well in the surface duct, as it is refracted upwards by the positive sound speed gradient and reflected downwards by the surface. Below this frequency, the sound waves will not fit, meaning these signals receive no benefit from the duct. For a particular mixed layer depth, there is a particular frequency that transmits best by surface duct.[19]

The depth of the mixed layer is not a fixed, unchanging value, however. The ocean plays host to internal waves that oscillate within the water column itself. The isothermal mixed layer and the thermocline below form an interface, where the water above and below have different densities. This causes gravity waves to propagate horizontally across the interface at the bottom of the mixed layer. When acousticians and aviators talk about a "layer," they are usually referring to this interface between the mixed layer and the thermocline below.

These internal waves cause the depth of the mixed layer to vary. This variation in mixed layer depth causes the size of the surface duct to change. As a result, sound propagation of waves at different frequencies tends to change. These fluctuations will impact sensor performance when the sensor and source are both in the mixed layer. Internal waves will also impact sound propagation across the mixed layer interface, when the source and receiver are on different sides of the layer.[20]

The Half Channel

When an isothermal water column forms, sound speed tends to increase from a minimum value at the surface to a maximum value at the seafloor. This condition results in a negative sound speed gradient that bends sound rays upwards. When the sound waves hit the surface they are reflected downwards where they are in turn refracted upwards again. This condition is known as a half channel. Half channel conditions are common in the Mediterranean Sea during the winter months and often form under the ice in the Arctic.

The Deep Sound Channel

In deep areas of the ocean, the typical configuration of the water column is an isothermal mixed layer near the surface with a thermocline and cold isothermal layer below. This condition produces a SSP with a maximum sound speed near the bottom of the mixed layer and a negative gradient below. Moving lower, sound speed decreases till a depth of several thousand feet, at which point it reaches a minimum value. Below this depth, the combination of a constant temperature and increasing pressure results in a positive sound speed gradient.

From **Figure 6**, we are already familiar with the type of deep water temperature profile and SSP that causes the DSC effect to occur. Note the near surface sound speed maximum and the sound speed minimum deeper in the ocean. We have annotated the sound speed minimum and its depth with a horizontal axis. Above this axis, the SSP will be negative, which tends to bend sound waves downwards. Below the minimum sound speed axis, we find a positive sound speed gradient, which refracts sound waves upwards. This temperature profile and SSP form the powerful ducting effect known as the DSC.

At short ranges in the DSC, sound spreads in a spherical manner. However, at long ranges, the ducting effect of the negative sound speed gradient above the minimum sound speed axis and the positive sound speed gradient below the axis cause sound spreading to more closely resemble cylindrical spreading. This causes a large reduction in TL.[21] Additionally, deep water absorption levels are much lower than those found in shallow water, meaning acoustic energy moving in the DSC enjoys very low rates of TL. The ducting effect and low absorption found in deep water combine to make the DSC an extremely effective long-range propagation path.[22]

The depth where the sound speed is lowest is known as the DSC axis. During the early years of the Cold War, the DSC was investigated extensively. Engineers determined that the DSC could be used to track ships and submarines at extremely long ranges, sometimes across entire ocean basins. Observations revealed that the DSC axis usually occurred at a depth of 3,000–4,000 feet.

Internal Sound Channels

In certain parts of the ocean, SSPs may lead to the formation of temporary and localized sound channels. These are known as internal sound channels. This type of propagation path is caused by a localized sound speed minimum that has a negative sound speed gradient above it and a positive sound speed gradient below it. Internal sound channels tend to form at a much shallower depths than the DSC, often with an axis approximately 300 feet below the surface.[23]

Internal sound channels are often found in the vicinity of strong ocean fronts.[24] For example, near a front or an eddy a region of cold, less saline water may intrude between a larger region of warmer, more saline water. This disparity in temperature and salinity causes a variation in the SSP and a localized duct to form.

Sound Arrival Paths

Now that we are familiar with the extended-range ducts and sound channels, we will discuss the different pathways that a sound wave can travel while moving between a submarine and a sensor. Each sound arrival path has different characteristics. Some are best used to generate long-range detection opportunities. Others are best suited for short-range tracking. The four arrival paths are: direct path, bottom bounce, convergence zone, and the RAP.

Direct Path

Direct path is the simplest type of sound arrival path. It is a short-range arrival path, where the sound wave follows an approximately straight line between the source and the receiver. There will not be any reflection off the surface or seafloor and only one change of direction due to refraction.[25]

Direct path is the most accurate sound arrival path and is the one that provides the best data for tracking and weapons employment. Traditionally, ASW aircrews have worked to convert whatever initial sound arrival path they detect into short-range direct path contact in order to plot the position of the submarine as accurately as possible.

Bottom Bounce

At slightly longer ranges than direct path, sound waves travel downwards from a submarine and reflect off the ocean bottom. These sound waves then move upwards and are detected by a sensor. This phenomenon is known as bottom bounce.

Bottom bounce is a challenging arrival path for aviators to utilize. This is partially due to the fact that there are several different pathways that sound waves can travel for the same sound source to receiver combination. This concept is known as acoustic reciprocity. The various arrival paths tend to cause constructive and destructive interference in the submarine signal. This leads to fluctuations in signal strength. Aircrews may detect a signal of interest arriving via bottom bounce and then suddenly lose the signal due to destructive interference.

Bottom bounce contact is also influenced by the effects that the seafloor itself has on the sound waves. As acoustic energy arrives at the seafloor, it will undergo scattering, reflection, and absorption. Some energy will be continue onwards in the direction of the wave front while some energy will be scattered in many directions. The tendency of energy to reflect cleanly and concentrate in a defined angle of departure is influenced by the roughness of the seafloor and the ratio between the acoustic impedance of the water and the bottom. A smooth bottom will cleanly reflect more sound in the original direction of travel while a rough bottom will tend to scatter sound waves. An ocean bottom that has a sharply different acoustic impedance compared with seawater will tend to sharply reflect sound off the bottom.

Under ideal conditions, with a smooth bottom and a sharp discontinuity between the acoustic impedance of the seawater and the seafloor, arriving sound waves will reflect sharply off the bottom with an angle mirroring the angle of arrival and with departing sound waves concentrated in this new direction. Acousticians refer to this phenomenon as a specular reflection. Unfortunately for ASW forces, specular reflection is not often observed in the ocean. Sound tends to scatter and reflect in a much more diffuse manner, leading to higher levels of TL.

Being able to properly characterize how much energy is lost to reflection and scattering during bottom loss interactions is very useful to ASW forces. Because of this, navies undertake surveys to characterize bottom losses in areas of tactical interest. Oceanographers try to determine how much energy will be lost overall and how different frequencies and angles of arrival affect these losses. While bottom losses are generally more pronounced at higher frequencies and larger angles of incidence, loss values are not uniform.[26]

When operating against a quiet submarine, an ASW aircraft may be unable to exploit the bottom bounce arrival path in deep water. Sonobuoys have very small hydrophones, and as such are only able to detect quiet targets over a range of several miles. When one considers that many abyssal plains are three to four miles deep, it becomes clear that sound waves propagating by bottom bounce must travel a long way before they reach a sonobuoy. Over this distance, the sound waves are likely to suffer high enough TL that they are not detectable once they reach a sonobuoy. As a result, aviators are often restricted to utilizing bottom bounce only in shallow waters.

Convergence Zone

In certain parts of the deep ocean, the SSP tends to focus sound rays into circular regions of very high signal strength. This leads to the formation of zones where a sensor can detect a target submarine at extremely long ranges. This phenomenon is known as the convergence zone path. It was used extensively by aircrews during the Cold War to locate their targets at long range and to localize to direct path contact and track.

In order for convergence zone contact to occur there must be a near surface sound speed maximum with a negative sound speed gradient below. The ocean must be deep enough that the sound speed can reach a minimum value and then increase in the deep ocean so that by the time it reaches the seafloor the sound speed is higher than the near surface maximum sound speed. An example of this type of SSP is shown in **Figure 9**.

Let us picture a sound source deeper than the mixed layer and therefore below the near surface maximum sound speed. As sound waves radiate outwards from the source, they are bent downwards by the negative gradient. As the sound waves move deeper into the water column, they are eventually bent back upwards by the positive sound speed gradient near the seafloor. These rays bend upwards and are focused in an area near the surface.

The area where these sound waves are focused is known as the reswept zone. The large amount of sound energy focused in the reswept zone causes much higher signal strength than would otherwise be possible given the long range from the target. When viewed from above, reswept zones form circular regions around the sound source. This leads oceanographers, acousticians, and aviators to refer to each reswept zone as an "annulus."

The distribution of sound energy across the reswept zone is not uniform. On the inner side of an annulus, many rays are focused together in a small area, leading to a very sharp rise in acoustic energy. This area is known as a caustic. Observations of caustics have found signal strength can rise on the order of 20 dB across a distance of only a few yards. The outer edge of an annulus, further away from the source, finds a much more diffuse distribution of acoustic energy. This difference between the inner and outer edge of a convergence zone annulus allow aviators to draw inferences about where the target may be located in relation to the sensor in contact.

The range from a sound source where a convergence zone may form and the width of the reswept zones vary widely. Their values are driven by water depth, surface temperature, SSP, and sound source depth.[27] The width of the reswept zone varies as well but is generally 5 to 10 percent of the range from the target.[28]

The relationship between the sound speed at the ocean bottom and the near surface maximum speed is critical in determining whether convergence zone transmission paths will form. For convergence zones to occur, the water must be sufficiently deep and the sound speed near the bottom must be sufficiently higher than the near surface maximum sound speed. In **Figure 9**, we observe a typical deep-water SSP capable of producing convergence zones. We refer to the depth where the sound speed achieves the near surface maximum value as the critical depth. The difference between the critical depth and the depth of the seafloor is known as depth excess. The difference between the sound speed at the bottom and that found at both the near surface maximum and at critical depth are known as sound speed excess.

The amount of depth excess and sound speed excess help predict how likely convergence zones are to form. **Table 1** shows the relationship between depth excess and the likelihood that convergence zone arrival paths will be available. We can see that larger amounts of depth excess and sound speed excess are correlated with higher chances of convergence zones formation.[29]

Table 1: Depth Excess versus Probability of Convergence Zone Formation

CZ Probability	Depth Excess	Sound Speed Excess
50%	1,200ft	22ft /sec
80%	1,800ft	33ft /sec

Source: "Reference Publication 33"

The SSP found in conditions that produce convergence zones generate refraction that tends to focus sound into reswept zones and raises the local SPL. However, as sound moves through the ocean, it still suffers TL. As a result, for aviators to exploit convergence zones, the target submarine must have radiated noise levels loud enough that even after TL is taken into account, there is still sufficiently high signal strength for detection to occur in the reswept zones.

ASW aircraft exploited convergence zones extensively during the Cold War. This was due in part to the high radiated noise levels of Soviet nuclear submarines. These searches often took place in deep water with sufficient depth and sound speed excess to support convergence zone formation. Improved quieting of nuclear submarines and the proliferation of quiet SSKs have lowered the radiated noise levels of nearly all modern submarines. Many smaller navies from developing nations operate in shallow waters, where convergence zone contact is not possible because of a lack of depth excess. As a result, convergence zone is still a possible arrival path for aviators to exploit. However, it much less prevalent in tactical scenarios today than it was several decades ago for the reasons listed above.[30]

Reliable Acoustic Path
Acousticians refer to direct path contact between a shallow source and a deep sensor or a deep source and a shallow sensor as the Reliable Acoustic Path, or RAP. This arrival path is referred to as "reliable" because it is not susceptible to near-surface effects or subject to bottom losses.[31] ASW forces take advantage of RAP in the intermediate ranges, beyond short-range direct path tracking but closer than convergence zone contact.

RAP has been the subject of renewed focus in the past two decades. As submarines became quieter in the 1980s and 1990s, convergence zone contact became less and less common and the ability of long-range cueing sensors to exploit the DSC decreased as well. This forced some ASW forces to shift from using small numbers of large fixed arrays for long-range detection and cueing and instead deploy networks of many small sensors deep in the ocean to exploit the RAP over intermediate ranges.[32]

Bathymetric Features

We have already alluded to the fact that the seafloor has a large effect on sound propagation. The ocean bottom reflects, scatters, and absorbs sound waves. It can channel and block acoustic energy. Let us examine the tactical implications of various bottom types and bathymetric features.

Close to the coast, the ocean bottom slopes gently away from the land, forming an area known as the Continental Shelf. This shelf may extend for only a short distance or for many dozens of miles. Eventually, the shelf drops away sharply from the surface, forming an area known as the Continental Slope. This sharp discontinuity from the Continental Shelf to the steep Continental Slope is known as the Shelf Break. Continental Shelves are generally relatively flat, although they may be interrupted in places by submarine canyons, ridges, and terraces.

The slope of the Continental Shelves and Continental Slopes vary depending on the overall bathymetry and features of the landmasses nearby. Wide, flat plains such as those found on the Atlantic Coast of North America fall away into a gently sloping Continental Slope. On the other hand, the jagged mountains of the Pacific Northwest of North America transition to a steep Continental Slope. Extreme geological features such as volcanic islands in the Pacific Ocean may find slopes that are as steep as 30 degrees.

In the deep ocean, we find wide abyssal plains. These plains are generally flat or slightly sloping. They are disrupted at certain points by ridges, rises, and seamounts. The centers of many ocean basins feature prominent submarine mountain ranges. Two examples of this phenomenon are the Mid-Atlantic Ridge, which bisects the Atlantic Ocean, and the Hawaiian-Emperor seamount chain. In certain places, the high peaks of these ridges climb above sea level. The Mid-Atlantic Ridge breaks sea level to form Ascension Island in the South Atlantic. In the Pacific, the above sea-level portions of the Hawaiian-Emperor seamount chain form the Hawaiian archipelago.

In most parts of the ocean, the seafloor is formed by several layers of material. The heterogeneous nature of the bottom makes absorption, scattering, and reflection phenomena difficult to predict and subject to a great deal of variability. The effect of various types of ocean bottom on sound propagation is an extraordinarily complex subject and has been written about at length.

Navies undertake bathymetric surveys to characterize the ocean bottom in various locations and measure the seafloor's effect on TL at various frequencies in each location. Aviators must be acutely aware of the type of ocean bottom in their operating area. They should review low-frequency and high-frequency bottom loss values in that area. If bottom losses levels are high, aviators may be much less likely to exploit bottom bounce arrival paths. On the other hand, low bottom loss values can help an aircrew predict and exploit bottom bounce during search and localization.

Tactical Exploitation of Bathymetry

The ocean bottom will influence sound propagation, making arrival paths such as bottom bounce or convergence zones more or less exploitable. It will also influence TL by funneling, spreading, or blocking sound waves. In this next section, we will discuss four tactical bathymetric phenomena: upslope enhancement, downslope enhancement, topographic shading, and topographic noise stripping. We will also discuss how a submarine may operate over or nearby underwater features such as seamounts or ridges to attempt to hide from ASW forces.

Upslope Enhancement

For either upslope or downslope enhancement to be possible, there must be a combination of appropriate oceanographic and bathymetric conditions and appropriately-placed sensors. In either case, the slope of the bottom must fall in a certain range of degrees. In the abyssal water beyond the Continental Slope, there must be an appropriate SSP and sufficient sound speed excess and depth excess to support convergence zone or DSC propagation.[33]

In the case of upslope enhancement, sound waves traveling via convergence zone or DSC propagation paths are funneled by the rising ocean bottom of the Continental Slope. The acoustic energy transitions from convergence zone or DSC propagation to bottom bounce propagation. As the sound waves move further from the source, they suffer from TL, but these losses are minimized as more acoustic energy is funneled together. For aviators to exploit upslope enhancement, their target must be in deep water, the boat must have sufficiently high radiated noise levels, and the sonobuoys must be in the shallow water of the Continental Shelf, inside the Shelf Break.

Downslope Enhancement

Downslope enhancement is the functional opposite of the upslope enhancement phenomenon. Here, a submarine in shallow water radiates noise that propagates via the bottom bounce path. As the Continental Slope drops away from the surface, the negative SSP below the mixed layer bends sound waves downwards. The sound waves transition from a bottom bounce to a convergence zone or DSC propagation path.[34] Aviators intent on exploiting downslope enhancement must be hunting a target in shallow water on the Continental Slope with a sensor located in deep water.[35]

Topographic Shading

During our discussion of the convergence zone and DSC phenomena, we mentioned that these propagation paths require a sufficient amount of depth excess. Without enough water below the critical depth, the deep positive sound speed gradient cannot bend sound waves upwards enough to channel them back towards the minimum sound speed axis and couple with the long-range propagation path.

Acousticians and ASW aviators use the term Topographic Shading to describe a situation where a bathymetric feature such as a ridge or seamount removes the required amount of depth excess from sound waves that would otherwise enjoy convergence zone or DSC propagation. From an acoustic perspective, the rays that enable long-range convergence zone or DSC propagation are "blocked" by the ocean bottom.

Up to this point, we have used the term critical depth to discuss the depth at which the sound speed equals the near surface maximum speed. In the case of Topographic Shading, using critical depth to calculate depth excess is not technically accurate. A more accurate treatment of the problem would find us plotting the sound speed at the target's depth.

For convergence zone or DSC propagation to occur, the target submarine is normally below the mixed layer depth. Examining the water column, we can find the location of near surface maximum sound speed. Moving deep into the water column, we find a corresponding speed, which we call critical depth. Moving shallower, we can plot the predicted depth of the target submarine and find the corresponding sound speed. Moving deep in the water again, we then can plot the sound speed that is identical to that at the target's depth. We call this measurement "conjugate depth." These conventions are shown in **Figure 9.**

When considering the effect of bathymetric features such as ridges or seamounts on convergence zone or SC propagation, aviators ought to reference conjugate depth rather than critical depth. If a bottom feature between the target and the sensor removes the required amount of depth excess, the sound waves will be scattered and absorbed by the seafloor and therefore unable to travel onwards to reach the sensor.

The orientation of the target submarine, the sensor, and the bathymetric feature drive whether or not topographic shading will interrupt convergence zone contact. To see why this is so, let us consider two scenarios. In the first, a sonobuoy is located one convergence zone away from a target submarine. Beneath the sonobuoy is a seamount that removes the required depth excess for convergence zone contact. In the second scenario, the sonobuoy is still one convergence zone range away from the target but the seamount is equidistant between the submarine and the sensor, at one-half convergence zone range.

In the first scenario, sound is bent downwards and then refracted upwards to form the reswept zone. While the seamount near the sonobuoy removes depth excess in this one location, there is no seafloor to restrict the ray paths from the target to the sensor. In the second scenario, sound is refracted downwards where it interacts with the seamount at half of the convergence zone range. The seafloor acts to reflect and scatter the sound. Comparing these two scenarios in **Figure 10**, we can clearly see how the orientation of target, sensor, and bathymetric feature is crucial to determining whether topographic shading will impact the tactical scenario and prevent long-range propagation phenomena.

Topographic Noise Stripping
With the proper type of Continental Slope and deep water that supports convergence zone arrival, aircrews can make use of the ocean bottom to reduce the effect of distant noises generated by surface ships. This reduction in AN allows the signal of a deep submarine source to stick out more, causing an artificial boost in the SNR, and helping an aircrew detect their target. For this technique, known as Topographic Noise Stripping (TNS), to work, the submarine must be below the mixed layer in deep water that supports convergence zone propagation. The sensor must also be located at an appropriate position on the Continental Slope.

As sound waves approach the Continental Slope, they move into an area of shallower water. Sound waves that originate from surface ships descend beneath the mixed layer and are refracted into a convergence zone propagation path if there is sufficient depth excess below the critical depth. Sound waves from the deep submarine are refracted into convergence zones as well if there is enough depth excess below the conjugate depth. Because the conjugate depth is shallower than the critical depth, the sound waves generated by the surface ship will run out of depth excess before the sound waves generated by the submarine do. **Figure 9** provides a helpful reference to visualize the relationship between critical depth and conjugate depth and why a surface vessel runs out depth excess before a submerged target.

As the ray paths from distant surface shipping run out of depth excess, they begin to hit the bottom and are attenuated through bottom losses and the transition to a bottom bounce propagation path. In this manner, distant AN is "stripped," while the sound waves from the submarine continue on via convergence zone propagation. This reduction in AN raises the SNR of the submarine signal, making detection easier.

Aviators and planners can determine where TNS is possible using SSPs for the deep water near their search area and contour charts of the ocean bottom. Finding the sound speed at the mixed layer, they can determine the critical depth and plot the locations on the contour charts where this depth occurs. Using assumptions about the likely depth of their target, they can determine a conjugate depth and add 1,200 feet to 1,800 feet to account to a good probability of convergence zone formation.[36] They can then plot the area on the Continental Shelf where this depth occurs on contour charts. The area between the two plotted lines shows the area where sensors may be placed to exploit TNS.[37]

Bathymetric Features and Submarine Evasion
Earlier in this chapter, we discussed the effect of a rough ocean bottom on sound propagation. A rougher bottom will more effectively scatter sound, leading to higher TL. In the same vein, features on the bottom like ridges and seamounts will block long-range propagation paths. Because undersea features tend to disrupt sound propagation, submarines are well served to seek them out.

There are many ways in which submarines may make use of bathymetry when trying to avoid detection. For example, sailing over or near a ridge or near a seamount tends to break up bottom bounce propagation paths. A submarine that denies the adversary bottom bounce detection opportunities forces the ASW force to rely on short-range direct path contact. This lowers the chance of detection during a search and hampers tracking operations. In another situation, a submarine might find itself attempting to avoid a fixed hydrophone array. Operating on the far side of a seamount or ridge can remove the required depth excess for convergence zone or DSC detection by the array, denying an ASW force these methods of long-range detection and cueing.

Fronts and Eddies

Oceanographers divide the ocean into large water masses that share many of the same characteristics. Classical ocean analysis divides water masses based on shared temperature and salinity, or "thermohaline," characteristics. Tactical oceanographers and aviators tend to classify water masses differently, breaking the ocean down into areas based on temperature.[38]

Between two different water masses, we find a discontinuity known as an ocean front. From a tactical perspective, a significant ocean front is an area where the sound propagation or TL is significantly altered. Aviators and planners usually detect and identify fronts by observing significant changes in the sea surface temperature, either from satellite measurements or observations with air-launched bathythermograph buoys. For aviators, fronts are easier to detect during the winter than during the summer. During the winter, solar heating is usually insufficient to cause significant changes in the thin surface layer, as opposed to the summer, when the surface layer is frequently subjected to diurnal temperature variations caused by solar radiation.[39]

Regardless of the method by which they are detected, ocean fronts can cause significant changes in sound propagation. Large changes in sound speed and mixed layer depth may be observed over very short distances. The interface of two different types of water often attracts large amounts of marine life, raising AN levels. The interaction of two water masses and the air above often causes rough surface waves, leading to changes in sea state and raising the AN.[40] In short, the environment is highly dynamic near an ocean front, making ASW very challenging.

The ocean also contains large rotating masses of water that are significantly colder or warmer than the water mass surrounding them. These areas are known as eddies, and usually form when a meandering portion of an ocean current separates from the larger current and spins off on its own. Eddies change how sound propagates compared to the water nearby and are therefore exploitable by both submarines and ASW aircraft.

Warm eddies tend to break off from a current and persist for six to eight months at a time, slowly transferring heat back to the cold water around them and eventually dissipating. The mixed layer and DSC axis are both deeper in a warm eddy compared to the cold water that surrounds the eddy. Aviators searching a warm eddy can expect improved surface ducts, better active sonar ranges, and improved conditions for the bottom bounce arrival path. However, strong negative sound speed gradients may form, bending sound downwards sharply. This sharply negative gradient acts to decrease direct path ranges, providing an advantage to a submarine. Because of this, boats may utilize warm eddies to loiter or to conduct noisy evolutions in an effort to decrease the chance of detection. Because of these reasons, submariners tend to view warm eddies as havens.[41]

Cold eddies are much longer-lived than warm eddies and may persist for up to two years. Cold eddies often have shallow mixed layers and generate poor surface ducting conditions. Aviators typically find weaker below layer sound speed gradients and a diffuse interface between the mixed layer and the thermocline. This weak layer allows sound to more easily move across the thermocline from the mixed layer to the deep water below. Cold eddies usually support improved convergence zone and DSC conditions compared with the water around them. The factors make cold eddies disadvantageous for submarines, with the exception of the complicated frontal areas near the eddy edges.[42]

Ambient Noise in the Ocean

AN has a very large effect on passive and active sonar performance. The background noise at particular frequencies of interest varies in its intensity both temporally and spatially. Luckily, AN values can be studied and predicted for particular locations and environmental conditions. These predictions allow ASW forces to estimate sensor performance and how likely they are to detect a target submarine in a given water column.

In this next section, we will briefly review the topic of AN and its tactical implications for aviators. We will first divide the AN spectrum into separate bands and discuss the characteristics of each band. We will then review the typical characteristics of AN both in the deep ocean and in shallow waters. We will discuss how aviators estimate AN values in operating areas. Last, we will summarize variable sources of AN, such as rain, biological sources, and seismic activity.

The Ambient Noise Spectrum

The oceanic AN spectrum can be divided into five bands. In each band, AN is caused by a different set of phenomena. In general, AN tends to decrease with increasing frequency. A typical AN plot of intensity versus frequency features a linear portion above 1 Hz with a distinct plateau in the region from 25 Hz to 100 Hz. The plot again becomes linear before marking a minimum value near 25,000 Hz and beginning to climb again. A typical oceanic AN plot may be found in **Figure 11**.

The sources of noise in Band 1 are largely unknown but are most likely due to tides, waves, and seismic activity. Hydrophones are affected by pressure and temperature variations. At the very low frequencies in Band 1, the change in static pressure due to tidal forces and temperature variations compresses and decompresses the hydrophone, making measures of SPL unreliable. Additionally, currents cause flow noise past hydrophones, which further complicates the process of accurately measuring AN at these very low frequencies.[43] Thankfully for aviators, sources in Band 1 are of no tactical interest to ASW forces, so the inability of acousticians to accurately measure in this region of the AN spectrum is no great obstacle.

Most AN occurring in Band 2 appears to be the result of oceanic turbulence. This turbulence causes pressure variations that register on hydrophones. Additionally, currents tend to shake or rattle hydrophones, causing AN in the process.[44] AN in Band 2 also shows a slight dependence on wind speed, with higher wind leading to higher AN levels.[45]

Unlike Bands 2 and 4, which show a linear relationship between higher frequencies and lower AN, Band 3 shows a distinct plateau. In this band, AN values hold relatively steady between 20 and 100 Hz. These values correspond with the dominant sound signature of surface ships. Distant shipping noise appears to be the primary origin of most AN in Band 3.

Band 4 noise is caused by phenomena near the ocean surface. The intensity of Band 4 noise is tightly correlated with wind speed, with higher winds speeds leading to higher AN values. Noise levels in Band 4 have been extensively studied and are often characterized and predicted using a series of working curves known as the Knudsen Spectra. By observing a wind speed or sea state in an operating area, aviators can use the Knudsen Spectra curves to predict the Band 4 AN levels at frequencies of interest.[46]

In Band 5, AN typically reaches a minimum intensity near 25,000 Hz. Above this frequency, AN intensity begins to rise with increasing frequency. AN in Band 5 is primarily "thermal noise" caused by the motion of seawater molecules.

Deep-Water Ambient Noise
In deep parts of the open ocean, AN values are relatively stable and predictable. Hydrophones will still register fluctuations, but the intensity of the fluctuations is small compared to those found in shallow water. Because the open ocean tends to be relatively homogeneous over large operating areas, ASW operators generally use a set of working curves to help estimate AN at frequencies of interest. These values allow planners to predict sensor performance.

The best-known set of working curves are the Wenz curves, named after a distinguished American acoustician. The Wenz curves combine representative linear values for AN in Band 2 and Band 5 along with variable curves based on shipping levels in Band 3 and wind speeds in Band 4. While using the curves, planners select an expected shipping level based on operational intelligence and a predicted wind speed based on meteorological forecasts. These individual curves are then faired together to produce a predicted AN spectrum for a particular location in the search area. **Figure 12** shows the Wenz curves.

Shallow-Water Ambient Noise
Typically, AN levels in shallow water are higher and more variable than those found in the deep ocean. A mixture of shipping activity, industrial noise,

and marine life combines to make shallow waters noisy and difficult places to conduct sonar operations. AN spectra vary a great deal from location to location and from time to time.[47]

Shallow water is not by its nature a noisier environment than deep water. In fact, shallow water that is free of man-made or biological activity generally shows wind-driven AN spectra very similar to those found in deep water.[48] However, shallow waters close to land tend to hold a large amount of noisy marine life. Close to the coast, a great deal of shipping and human activity tends to ensonify the water. Shallow-water AN spectra tend to vary much more in the lower frequencies. High levels of oceanic turbulence and variations in shipping levels tend to drive large fluctuations in Band 3 noise while the higher frequency Band 4 noise caused by wind action will generally not vary as greatly.[49]

AN in coastal areas and harbors shows very high variability with respect to time. Tides and their associated currents produce turbulence that raises Band 3 noise levels. Industrial noise in harbors is tied to cycles of human activity. For example, AN levels are generally higher during the day and lower at night because of commercial activity. In a similar way, harbors tend to be noisier during weekdays and quieter during weekends.[50]

Variable Ambient Noise Sources
In the previous sections, we discussed the major factors that influence AN levels in the various bands of the spectrum. There are additional transient sound sources that cause additional AN in the ocean. The main variable noise sources are rain, explosions, geologic activity, and biological sources.

Rain striking the ocean surface can increase local AN levels. The intensity of noise caused by rain is proportional to the intensity of the rain and is slightly correlated with the size of the area over which the rain is falling. AN caused by rain generally spans from 1,000 to 10,000 Hz and varies in strength depending on the intensity of the rainfall. AN caused by intermittent and moderately heavy rain shows a decrease of intensity with increasing frequency. Heavy rain, on the other hand, tends to produce relatively constant AN increases across the frequency spectrum, leading to a profile more akin to "white noise."[51]

Companies engaged in offshore oil and gas extraction often use explosive sound sources to map the ocean and the geological structure beneath the ocean bottom. As oceanic resource extraction has become more common over the years, AN caused by explosions has increased.

Seaquakes are a source of transient low-frequency AN. Most quakes are small and generally last only a few seconds. However, seismically active regions of the ocean such as the Mid-Atlantic Ridge or East Pacific Rise witness a near-continuous set of small, individual seaquakes. Taken together, these quakes increase AN levels. Active volcanoes also produce low-frequency noise, adding another geological noise source to the AN spectrum.

Three types of animals produce sound in the ocean. They are crustaceans, croaking fish, and cetaceans. In shallow, coastal waters with rock, shell, or weed-covered bottoms are home to large numbers of snapping shrimp. These small creatures snap their claws together, creating a broad spectrum of noise between 500 and 20,000 Hz.[52] Croaking fish drum muscles on their air bladders, producing a tapping sound similar to that of a woodpecker hammering on a tree. Cetaceans such as whales, dolphins, and porpoises force air through their larynxes, creating a complicated set of squeals, barks, whistles, and songs. Biological sound sources are easily identified by sonar operators. Despite being easy to identify, these sounds still increase AN levels, degrading passive sonar performance.

Oceanic Clarity and Turbidity

Depending on the clarity of seawater, the ocean may either hide a submerged submarine or allow it to be detected visually while sailing at a shallow depth. As a result, it is useful to aviators to know how clear or opaque the seawater in their operating area is likely to be. Clarity is the measurement of how transparent water is to light and therefore how possible it will be to locate a submerged object. Turbidity is an optical determination of water clarity.

In general, seawater is less turbid than fresh water. Salt ions in seawater bond with and collect suspended particles. This interaction increases the weight of the suspended particles, which in turn causes them to sink to the ocean bottom. In general, coastal waters tend to be more turbid than the open ocean. The exception is some tropical areas where the water is extremely clear, even when very close to land.

Radar Performance

Radar propagation is affected by the atmosphere in a manner similar to how sound propagation is affected by the ocean. Just as sound waves are refracted in the water, so too are radio waves refracted by air. The ocean surface reflects radio waves just as the seafloor reflects sound waves. Variations in the temperature and humidity profile in the atmosphere cause radar ducting phenomena to occur just as variations in temperature and pressure cause sound ducting to occur in the water column.

As altitude increases, temperature and pressure generally decrease. This causes a variation in air density and a resulting change in the speed of light. This variation in the speed of light and the demands of Snell's Law compels radio waves to bend downwards as they radiate outwards. Depending on a measure known as the refraction index, radio waves will bend at a lower rate than the curvature of the earth (sub-refraction), the same rate as the curvature of the search (standard refraction), or a higher rate than the curvature of the earth (super refraction).

Radio refractivity depends mostly on the air's humidity profile. As a result, when temperature increases with height or water vapor decreases unusually rapidly with height, ducts that trap radar energy may form. Just as ducts tend to funnel sound waves, so too does atmospheric ducting cause radar waves to travel further. Surface ducts caused by evaporation trap radio waves and carry them further than the curvature of the earth would suggest is possible. Elevated ducts act to trap energy, making a patrol aircraft's radar detectable by a submarine's ESM equipment at longer-than-normal ranges.

Radar performance is also affected by sea states. Wave action tends to produce clutter on a radar scope. This requires engineers to devise processing techniques to pick out a small target, such as a periscope or snorkel, from the wave action that surrounds the object. Higher sea states will generally reduce radar performance, leading to shorter detection ranges or rendering the radar completely ineffective.

Sea State Predictions

Sea state has a great influence on ASW aviators. First, a high sea state will degrade radar performance. Second, large waves may cause a phenomenon known as "buoy washover," where a large wave or crashing whitecap submerges the radio antenna that links a sonobuoy with the aircraft overhead. This disruption breaks contact with the sonobuoy and interrupts target tracking.

High sea states are usually correlated with high wind speeds. A strong wind moving over a rough sea will usually generate considerable mechanical turbulence in the air above. This results in a very rough ride for aviators nearby. Moderate and heavy turbulence are quite uncomfortable and frequently lead to aviators becoming airsick, as they are thrown around while concentrating intensely on their equipment. Sick crewmembers rapidly cause an aircrew to lose its tactical effectiveness.

Atmospheric Visibility

Meteorological visibility refers to the transparency of air. Light moving between an object and an observer is scattered and absorbed by gases and particulate matter in the atmosphere. The ability to detect an object visually is very important to an aircrew, especially if they are intent on conducting a visual search for a submarine. Forecasting visibility before a search also helps aviators predict how well they will be able see, identify, and avoid surface ships, oil platforms, and aircraft in their operating area.

In general, as humidity levels increase, atmospheric visibility tends to decrease. Water-absorbing particles take on more water at high relative humidity levels. This allows them to absorb more light and in turn reduces visibility. Atmospheric visibility is also reduced by particulate matter, which tends to scatter light. Scattering phenomena have different effects depending on the sensor in use. For example, fog and haze are caused by water droplets suspended in the air. These particles are very effective at degrading the performance of the human eye because they scatter a great deal of light in the visible spectrum. A sensor that operates in the IR spectrum will be less affected by fog or haze because IR radiation is of a different wavelength than visible light and is therefore scattered and absorbed at lower rates by fog and haze.

Notes

1. "Reference Publication (RP) 33: Fleet Oceanographic and Acoustic Reference Manual" (Stennis Space Center, MS: U.S. Navy Naval Oceanographic Office, April 1999), p. 12.
2. Robert J. Urick, *Principles of Underwater Sound* (Westport, CT: Peninsula Press, 1983), p. 100.
3. Ibid., pp. 104–05.
4. Seawater suffers absorption at much higher rates than fresh water or distilled water, due to presence of dissolved salts. Magnesium sulfate ($MgSO_4$) is the largest contributor to absorption in seawater, despite its relatively low levels in the ocean compared to sodium chloride (NaCl). At higher frequencies, Boric acid also plays a role in ionic relaxation losses. Daniel R. Raichel, *The Science and Applications of Acoustics, Second Edition* (New York: Springer, 2006), p. 419.
5. Urick, *Principles of Underwater Sound*, pp. 108–10.
6. "Reference Publication 33," p. 14.
7. Urick, *Principles of Underwater Sound*, p. 255.
8. Ibid.
9. Ibid., p. 256.
10. Acousticians use a measurement known as the Rayleigh parameter to model the roughness of a surface for wave reflection. The Rayleigh parameter, or "R value," takes into account the size of the wave and angle of arrival. For a large R value, the surface will tend to act as a reflector. For a small R value, the surface will tend to scatter more of the acoustic energy. Ibid., p. 129.

11. Ibid., pp. 139–40.
12. Assuming a set salinity value is often suitable for ASW operations but does not accurately represent certain water columns. Areas close to river mouths frequently see very large variations in salinity values even over small distances. Many parts of the ocean show little variation in salinity values from the surface to depth whereas others vary a great deal. For example, the Strait of Gibraltar finds warmer, more saline water from the Mediterranean Sea interfacing with colder, less saline water from the Atlantic Ocean. The heavier, more saline water from the Mediterranean sinks towards the bottom and flows out into the Atlantic basin. The lighter, less saline water from the Atlantic flows into the Mediterranean creating a strong, fast-moving current that moves from west to east. Suitable knowledge of a search area is the best tool an aircrew has to decide if their assumptions about salinity levels are realistic.
13. Readers may commonly find Sound Speed Profile referred to as Sound Velocity Profile, or SVP, in older works. Because acousticians are measuring sound speed, a scalar quantity, it is incorrect to talk about sound speed measurements as velocities, since velocities are vectors. The use of the term SSP is more technically correct and is more frequently used in modern literature.
14. The surface layer is generally thin and not particularly relevant to ASW aircraft, as air-launched sonobuoys usually deploy hydrophones anywhere from 50 to 1,000 feet below the surface. The surface layer may, however, drastically affect hull-mounted sonars on surface ships. Solar radiation tends to heat the surface layer during the day, leading to a temperature rise in the afternoon. This increase in temperature near the surface bends sound waves sharply downwards, degrading the performance of hull-mounted active sonars. This phenomenon is known as the "afternoon effect." "Reference Publication 33," p. 79.
15. Urick, *Principles of Underwater Sound*, p. 117.
16. Half channel ducts are common during the winter months in the Mediterranean Sea and are nearly always present under ice shelves in the Arctic. "Reference Publication 33," p. 80.
17. Ibid., pp. 75–6.
18. Urick, *Principles of Underwater Sound*, p. 150.
19. Ibid., p. 151.
20. Ibid., p. 153.
21. Ibid., pp. 162–3.
22. Ibid., pp. 110–11.
23. Ibid., p. 169.
24. "Reference Publication 33," p. 83.
25. Ibid., p. 74.
26. Ibid., p. 17.
27. Ibid., p. 91.
28. Due to the complexity of the factors affecting reswept zone size, the easiest way for aviators to calculate reswept zone width is to use a TL plot. Using predicted environmental values, planners can calculate a FOM for the target and water column. They can then plot the FOM and see where it intercepts the local peaks in the TL plot for each convergence zone. The point where the FOM plot crosses the inner and outer portions of the TL plot show the inner and outer edges of each annulus. The difference between each range can be used to define the width of the reswept zone. Ibid., p. 92.
29. Aviators use two rules of thumb to determine whether convergence zones are likely based on observations from environmental predictions or bathythermograph sonobuoys deployed during a search. A depth excess of 1,200 feet or a sound speed excess of 22 feet per second will result in a 50 percent probability of convergence zones forming. A depth

excess of 1,800 feet or a sound speed excess of 33 feet per second will result in an 80 percent probability of convergence zone formation. These assumptions are based on a suitably loud target submarine and the absence of bathymetric features like seamounts or ridges that would disrupt the sound arrival paths. Ibid., p. 90.

30. Aviators should keep in mind that SE is the key driver of whether convergence zone contact can be exploited. Sonobuoys are cheap and small sensors, meaning they are limited in their ability to gather sound in and increase the SNR. While a sonobuoy may not generate sufficient SE to detect a submarine by convergence zone, another sensor in the same water column might be able to detect this arrival path. A towed sonar array being employed by a submarine or a surface ship may be able to achieve convergence zone contact and serve as a cueing source for an ASW aircraft. If the water column supports convergence zone formation because of the SSP and sufficient depth and sound speed excess, aviators should be aware other cueing sensors may utilize this phenomenon even if their own sensors cannot generate convergence zone detection opportunities.

31. Urick, *Principles of Underwater Sound*, p. 195.

32. The U.S. Navy developed a seafloor-based sensor network program known as FDS during the 1990s. The system utilized a large number of arrays to exploit RAP. Late stage operational test and evaluation was canceled in 1996. Lt. John Howard, USN, "Fixed Sonar Systems: The History and Future of the Underwater Silent Sentinel," *The Submarine Review* (April 2011), p. 11.

33. "Reference Publication 33," pp. 98–9.

34. It is thought that some of the AN in deep water is due in part to essentially this phenomenon. Sound produced by shipping in coastal waters is refracted downwards off the Continental Slope and enters the DSC. This sound manifests itself as low-frequency AN in the same band as most surface shipping. Urick, *Principles of Underwater Sound*, p. 215.

35. "Reference Publication 33," p. 99.

36. Recall from our previous discussion that 1,200 feet of depth excess will usually produce a 50 percent probability of convergence zone formation. 1,800 feet of depth excess will usually result in an 80 percent probability of convergence zone arrival path formation.

37. "Reference Publication 33," pp. 102–04.

38. This is due in part to the fact it is far easier for tactical ASW forces to measure temperature rather than salinity. Without specialized, and therefore expensive, survey equipment, it is difficult to directly measure salinity.

39. "Reference Publication 33," p. 42.

40. Ibid., p. 47.

41. Ibid., p. 51.

42. Ibid.

43. Urick, *Principles of Underwater Sound*, p. 209.

44. Ibid., pp. 205–06.

45. Ibid., p. 209.

46. The Knudsen Spectra take their name from Vern Knudsen, an American physicist who oversaw the compilation and publication of various acoustic surveys undertaken during World War II. It is interesting to note that much of wartime survey data was collected in support of developing acoustic triggering mechanisms for mines.

47. Urick, *Principles of Underwater Sound*, pp. 212–13.

48. Ibid., p. 213.

49. Ibid., p. 216.

50. A similar relationship between shipping-driven AN and overall AN levels holds true when considering cycles of economic activity. Worldwide economic growth has caused AN levels in the open ocean to grow at a rate of about 3.3 dB per decade from 1950

to the present day. Strong economies result in greater trade and higher amounts of shipping traffic. The economic downturn of 2008 saw a measurable reduction in oceanic AN because more merchant ships were sitting idle rather than steaming underway and making noise. George V. Frisk, "Noisenomics: The Relationship Between Ambient Noise Levels in the sea and Global Economic Trends," *Scientific Reports*, Vol. 2, No. 437 (June 2012).

51. Urick, *Principles of Underwater Sound*, p. 219.
52. Ibid., p. 217.

Chapter 9

Sound Propagation Analysis

While planning a mission, ASW aviators always attempt to calculate how far away from a sensor they can detect a target submarine. As we have seen in the previous chapter, sound propagation in the ocean is highly complex. Propagation becomes even more complicated when we take into account the fact that the target submarine is moving relative to different water masses and bathymetric features and that the boat may vary its radiated noise by changing its operational posture. Modeling sound propagation is critical to help aviators determine where and how they should employ their sonobuoys to detect and track the target.

Analyzing Transmission Loss (TL) allows ASW planners to calculate the detection range of their sensors. By modeling sound refraction and reflection, they can determine the various ray paths that sound waves are likely to take as they travel from a target submarine to a sensor. Calculating TL and detection range assists aviators in many tactical tasks, allowing them to do the following:

- Determine detection ranges for search planning
- Generate measures of effectiveness for search operations
- Determine ideal sonobuoy drop points and sensor depth settings
- Identify tactically relevant sound arrival paths and forecast possible long-range propagation phenomena
- Determine appropriate localization patterns and geometry
- Employ efficient sonobuoy tracking patterns
- Predict and optimize LWT seeker performance

Accurately calculating TL and detection range is not always simple. In order for the calculations to be accurate, aviators must predict the environmental conditions in the ocean, utilize appropriate source frequencies and source levels that match the sounds emitted by the submarine, and predict the submarine's expected tactics. If the aviators use incorrect values for these various parameters, the TL and detection range calculations will be inaccurate.

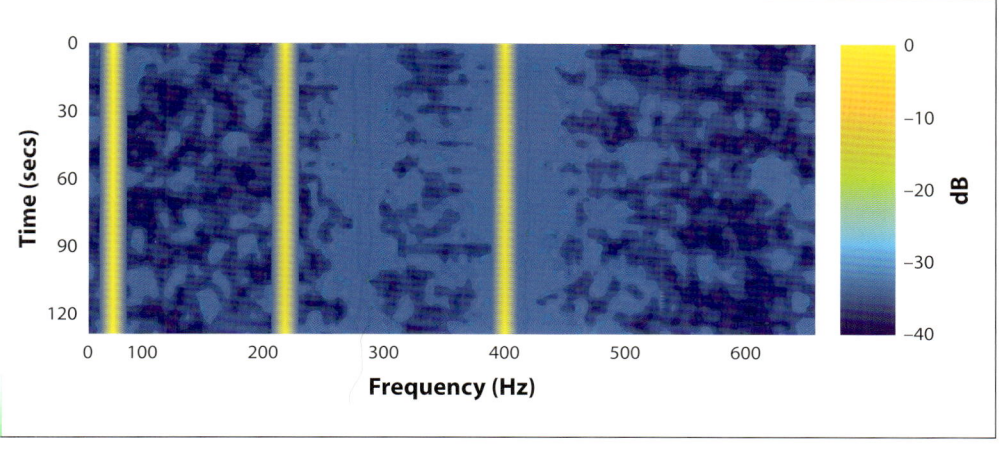

Figure 1: Notional LOFARGram.

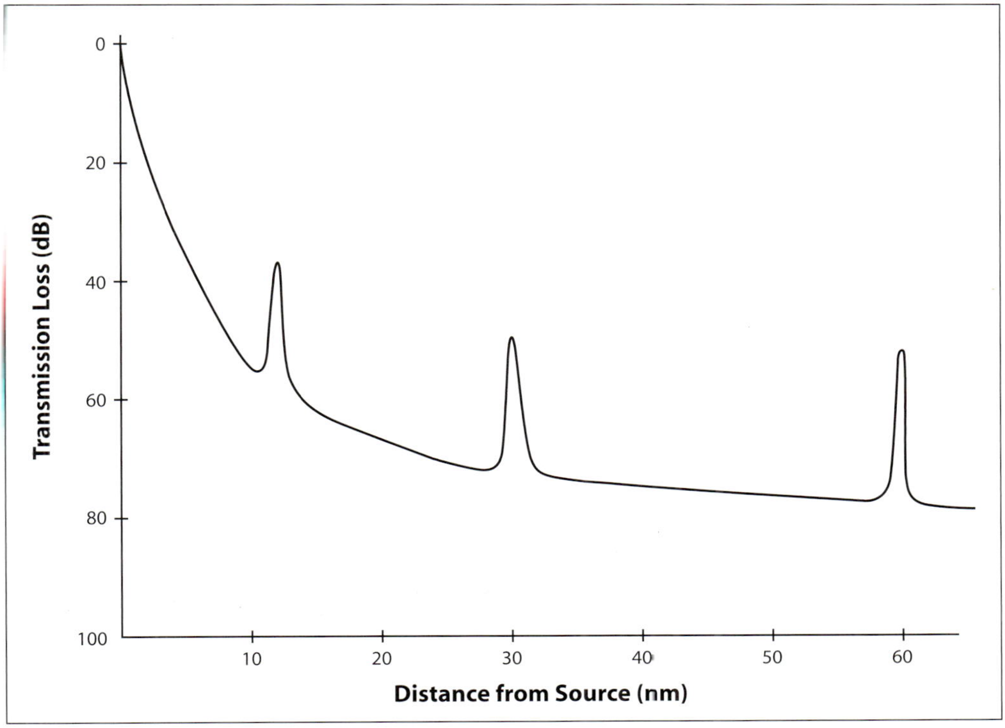

Figure 2: Transmission Loss Curve.

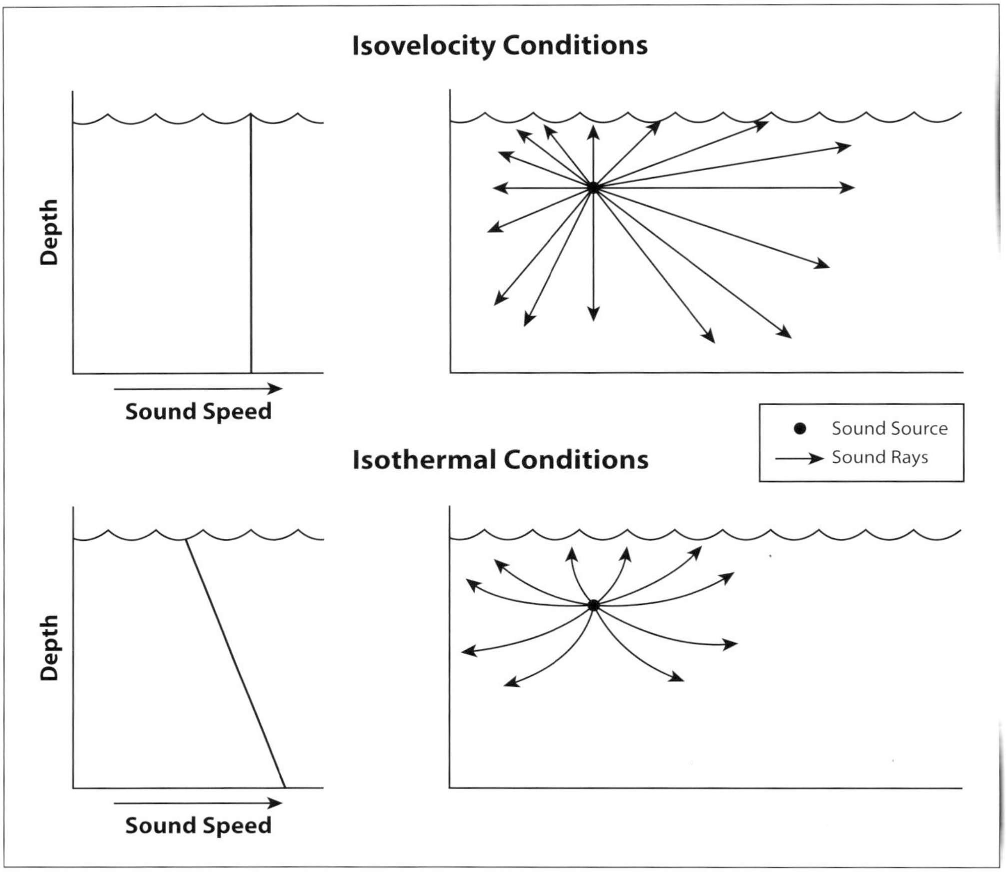

Isovelocity Conditions

Depth

Sound Speed

Sound Source
Sound Rays

Isothermal Conditions

Depth

Sound Speed

Figure 3: Sound Propagation under Isovelocity and Isothermal Conditions.

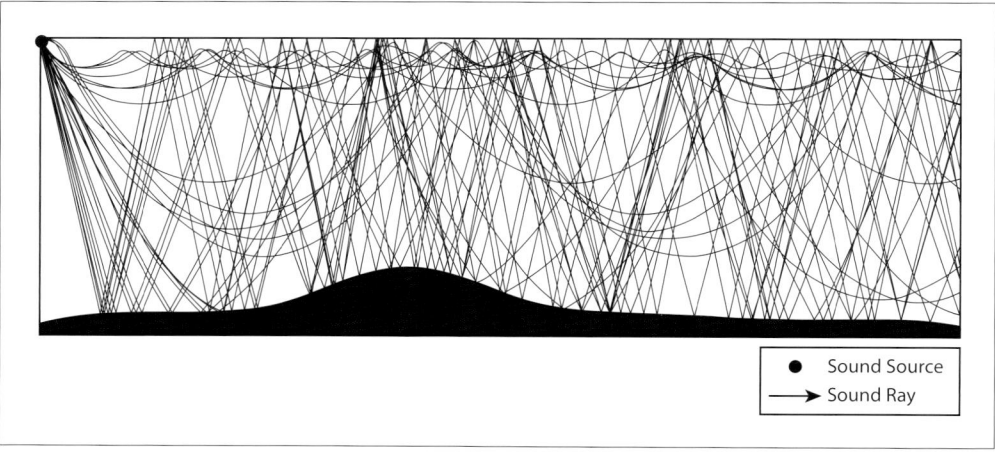

Sound Source
Sound Ray

Figure 4: Ray Trace of Sound Propagation.

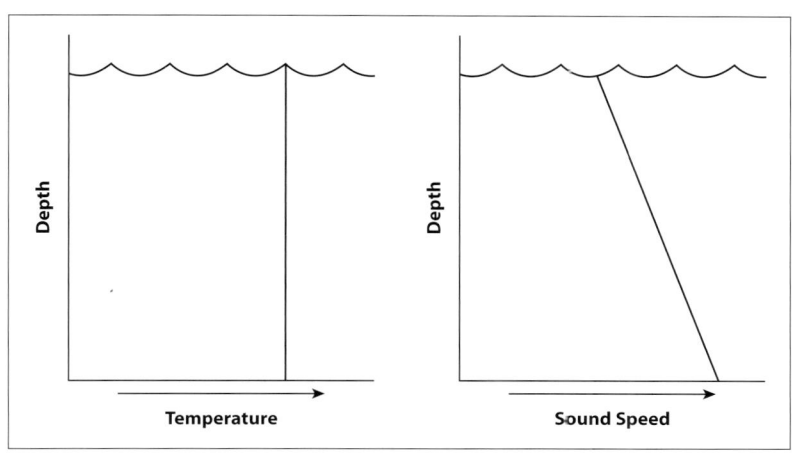

Figure 5: Shallow Water Isothermal Water Column.

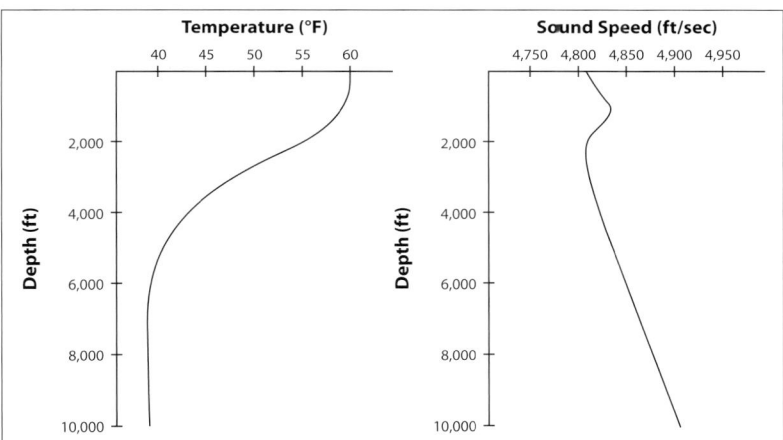

Figure 6: Deep Water Temperature Versus Depth Plot and Sound Speed Profile.

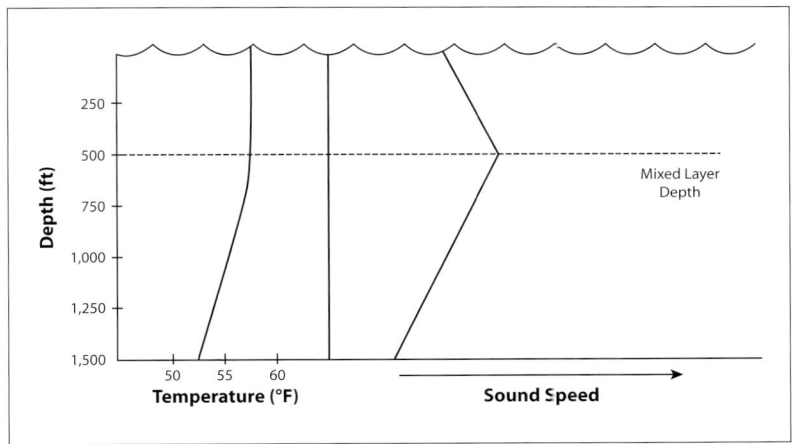

Figure 7: Mixed Layer Temperature Versus Depth Plot and Sound Speed Profile.

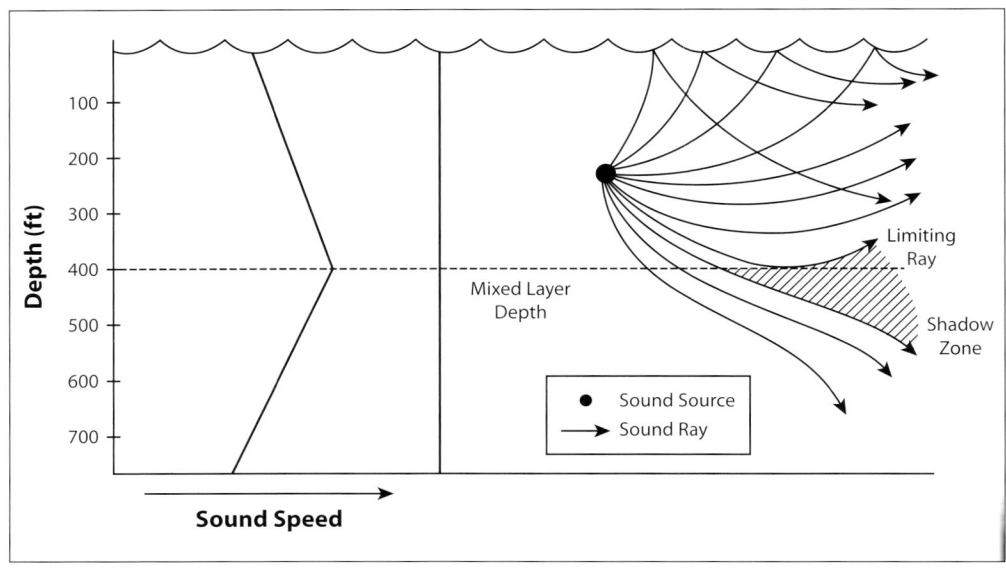

Figure 8: Surface Duct, Limiting Ray, and Shadow Zone.

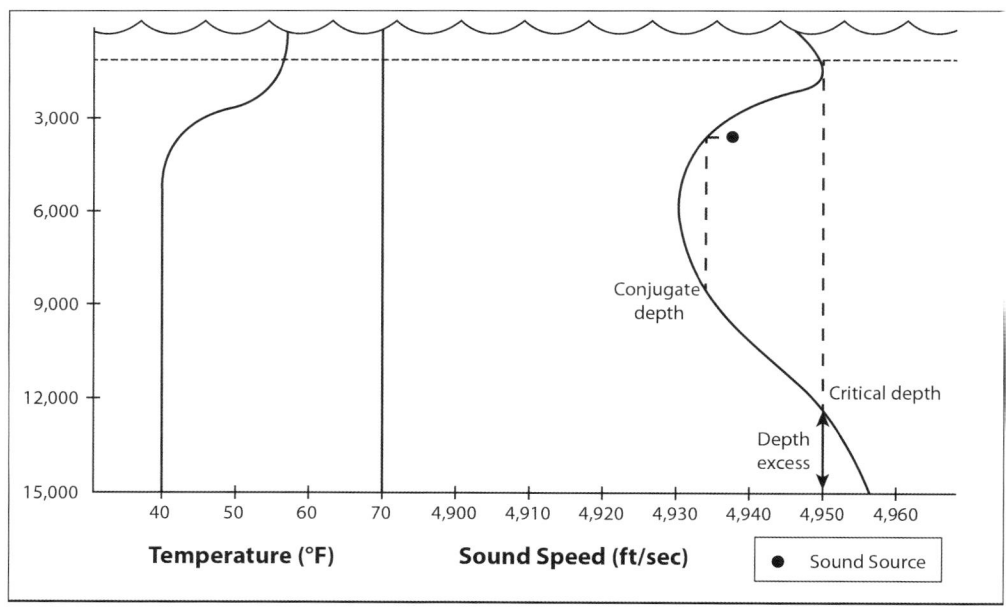

Figure 9: Convergence Zone Propagation, Critical Depth, and Conjugate Depth.

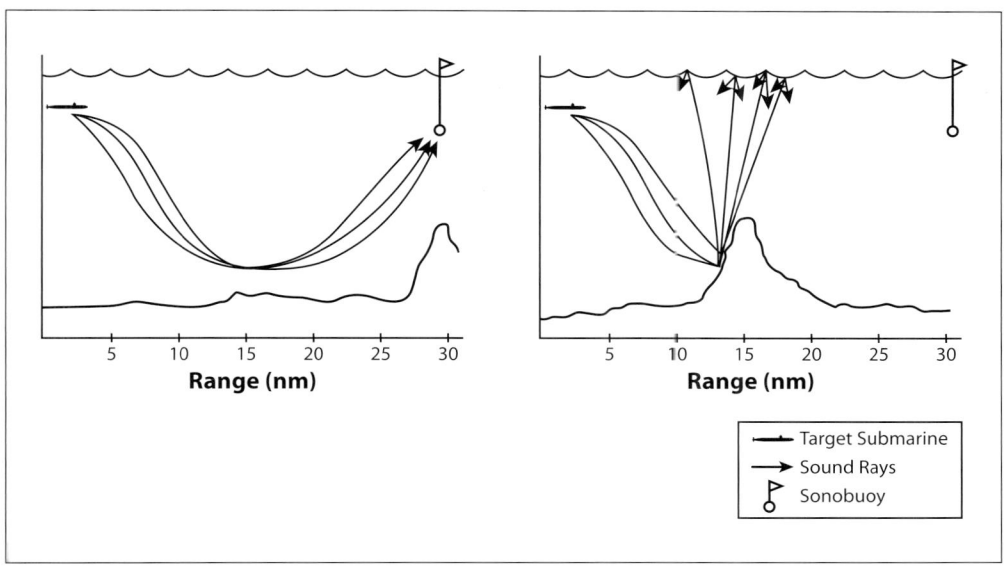

Figure 10: Bathymetric Effect on Propagation Path Availability.

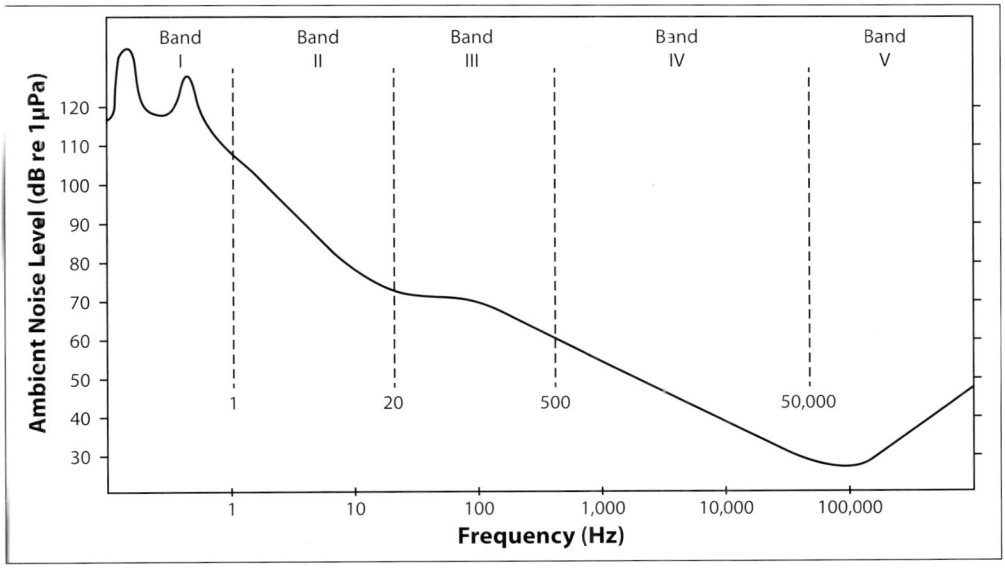

Figure 11: Ambient Noise in the Ocean.

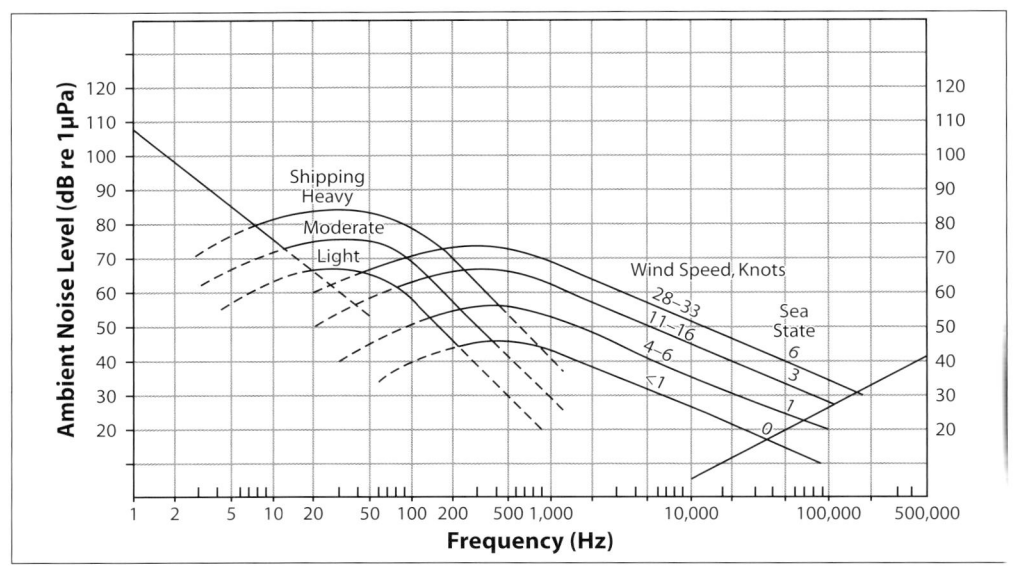

Figure 12: Wenz Curves.

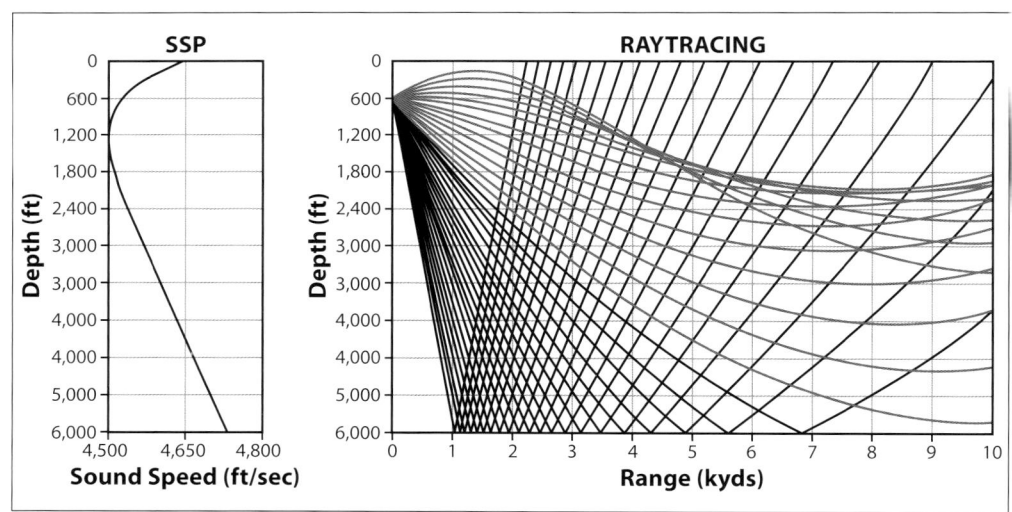

Figure 13: Sound Speed Profile and Ray Trace.

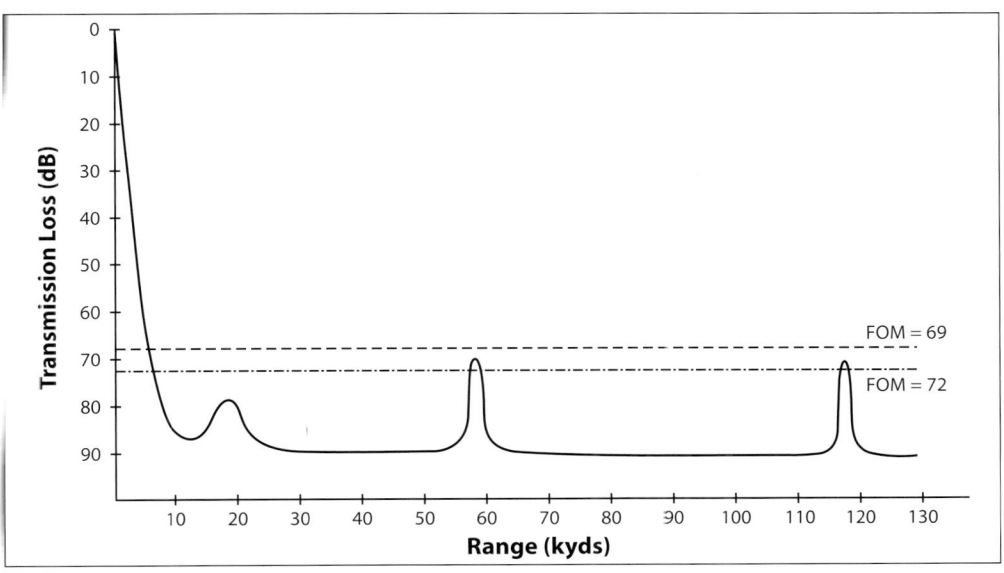

Figure 14: Transmission Loss, Figure of Merit, and Ambient Noise Impact on Sensor Performance.

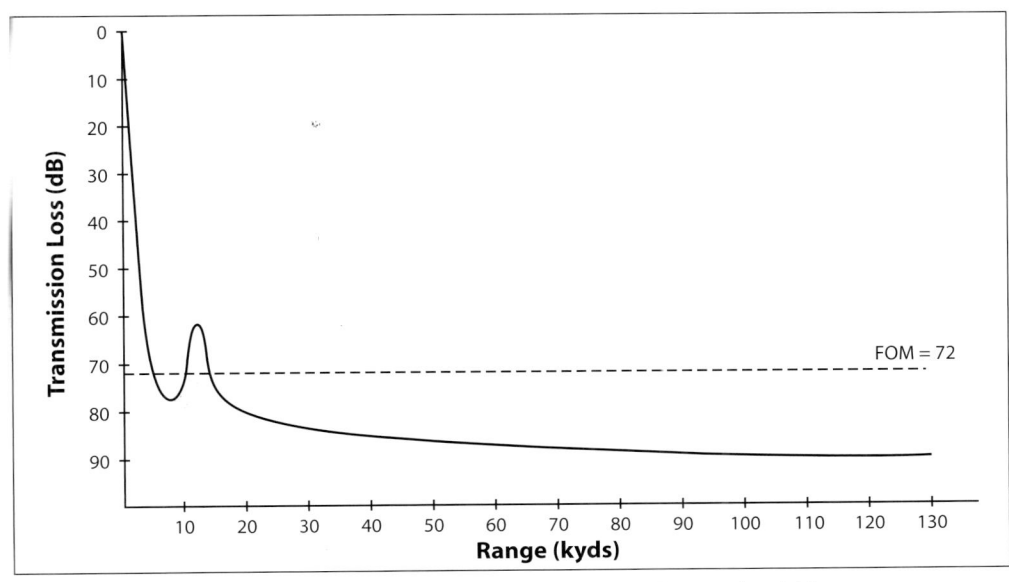

Figure 15: Transmission Loss Plot with Direct Path and Bottom Bounce Available.

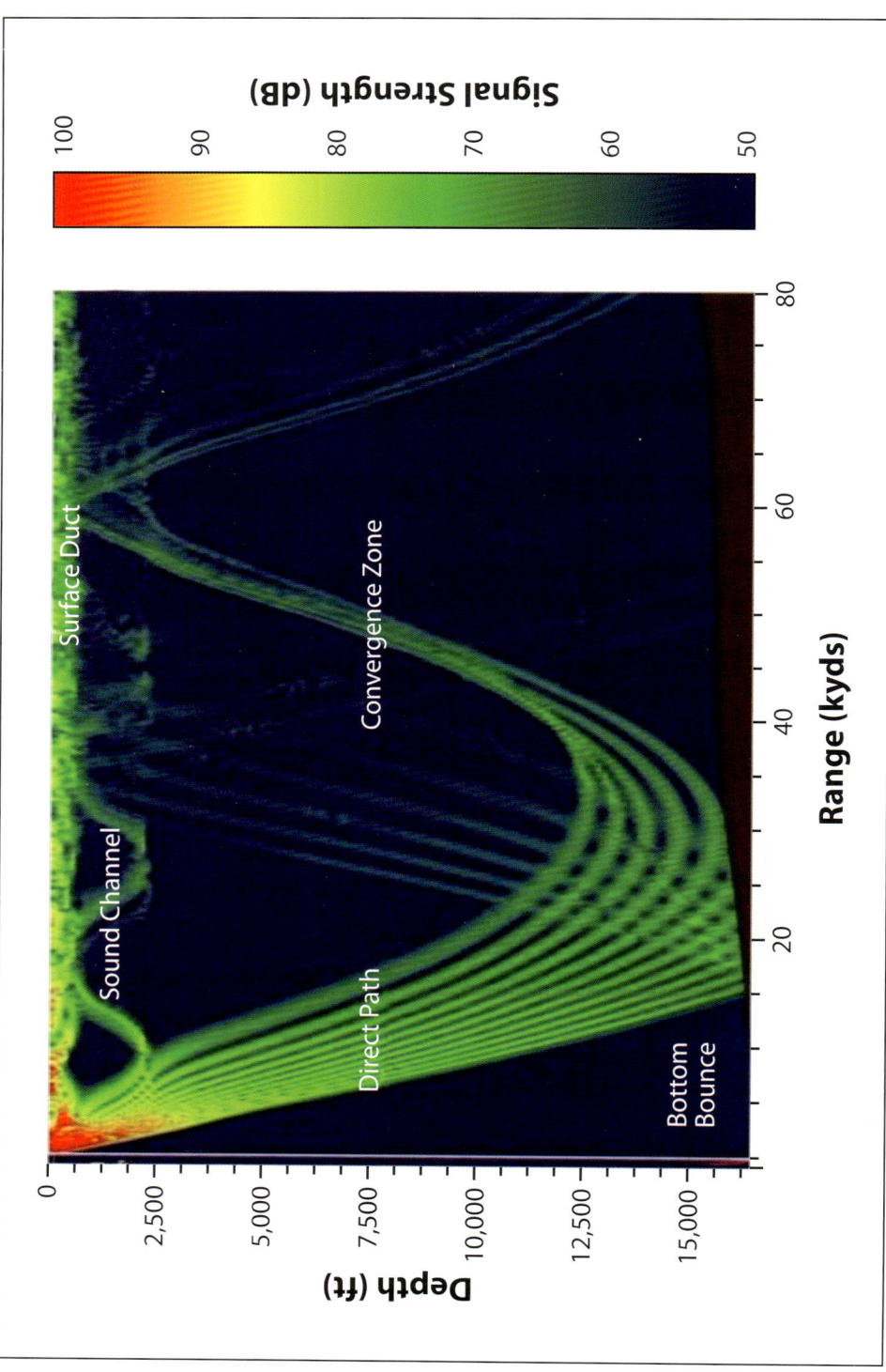

Figure 16: Notional Full Field Plot.

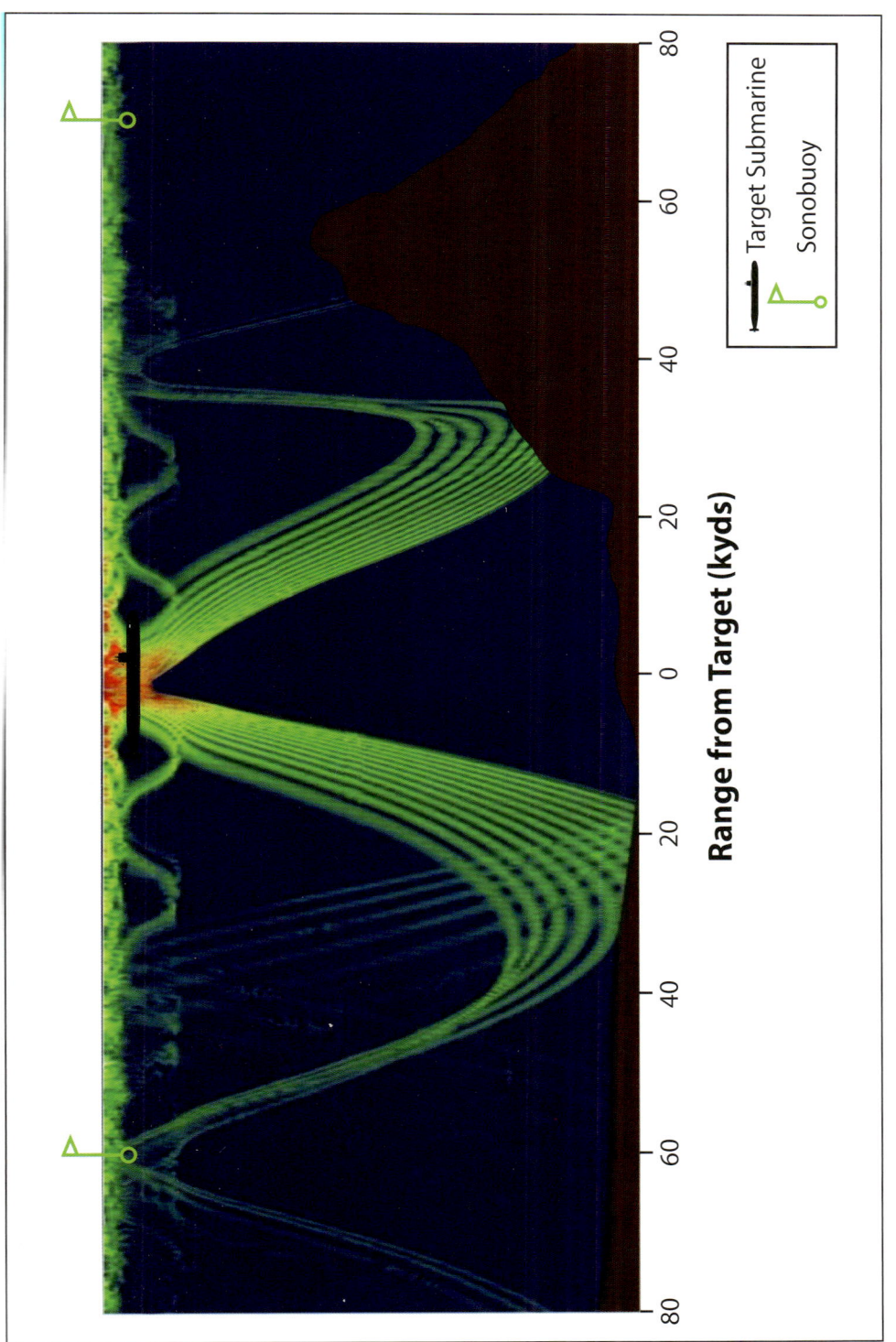

Figure 17: Bathymetric Impact on Full Field Plot and Convergence Zone Availability.

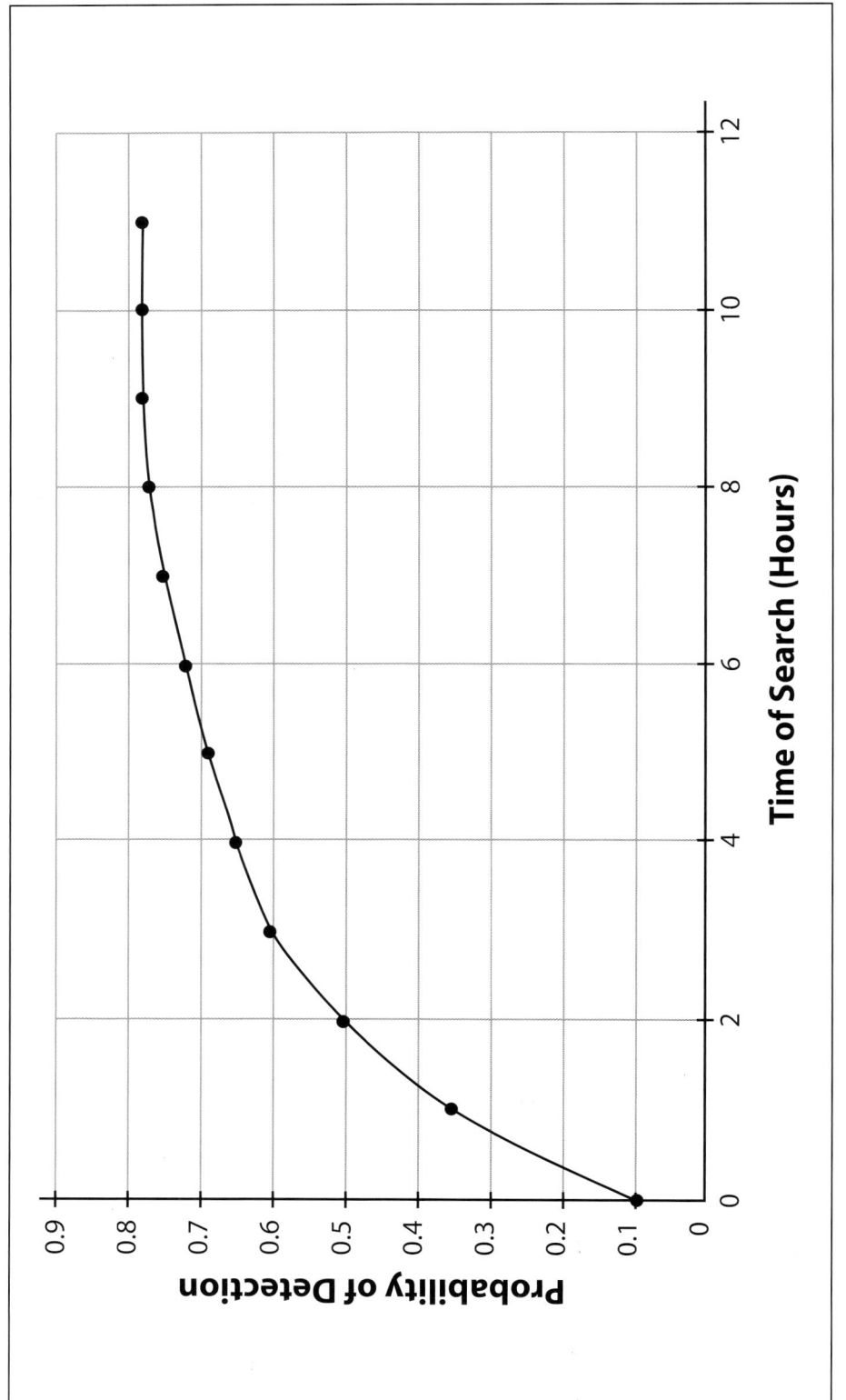

Figure 18: Cumulative Detection Probability Versus Time.

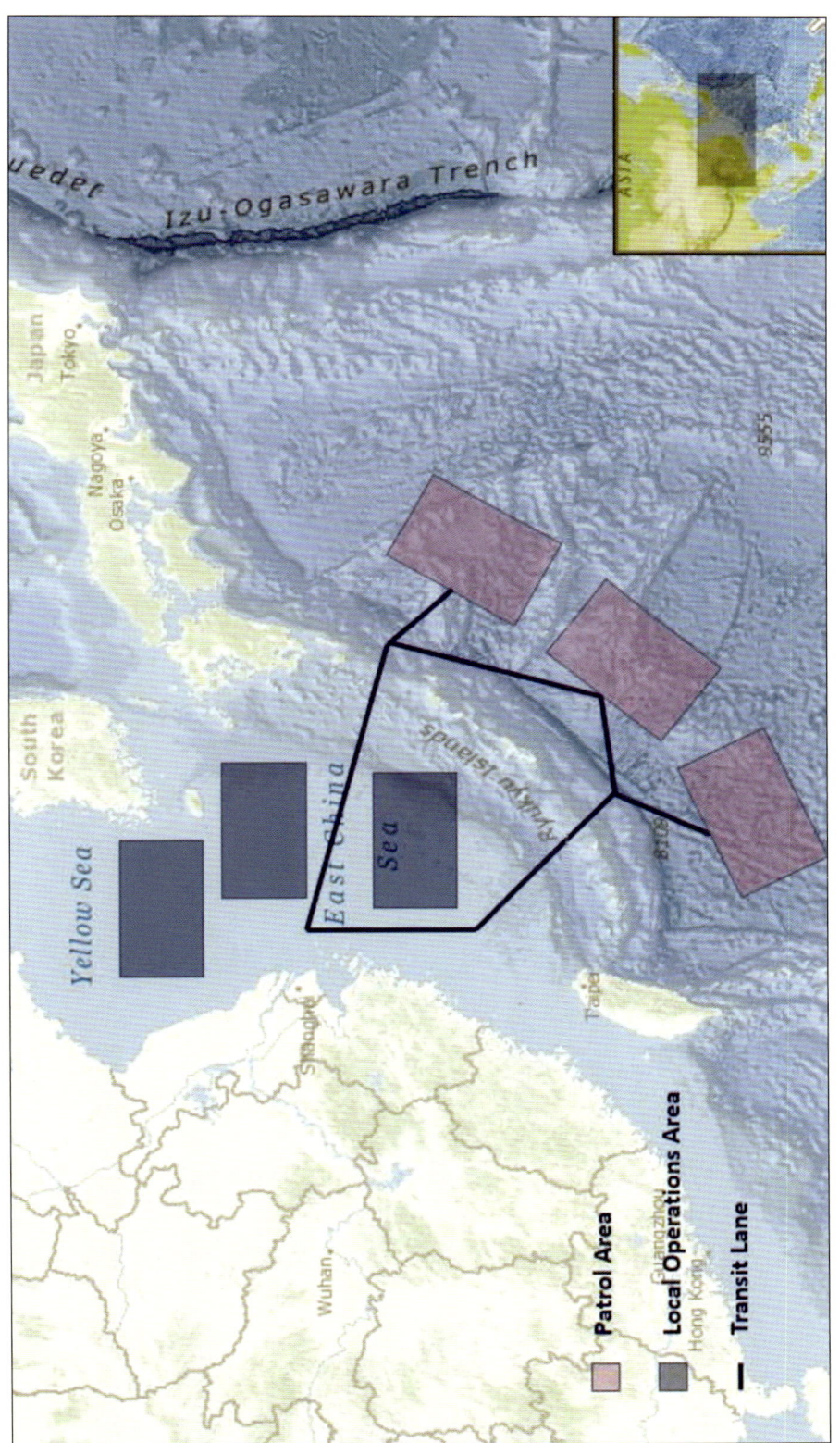

Figure 19: Notional Local Operating Areas, Transit Lanes, and Patrol Areas.

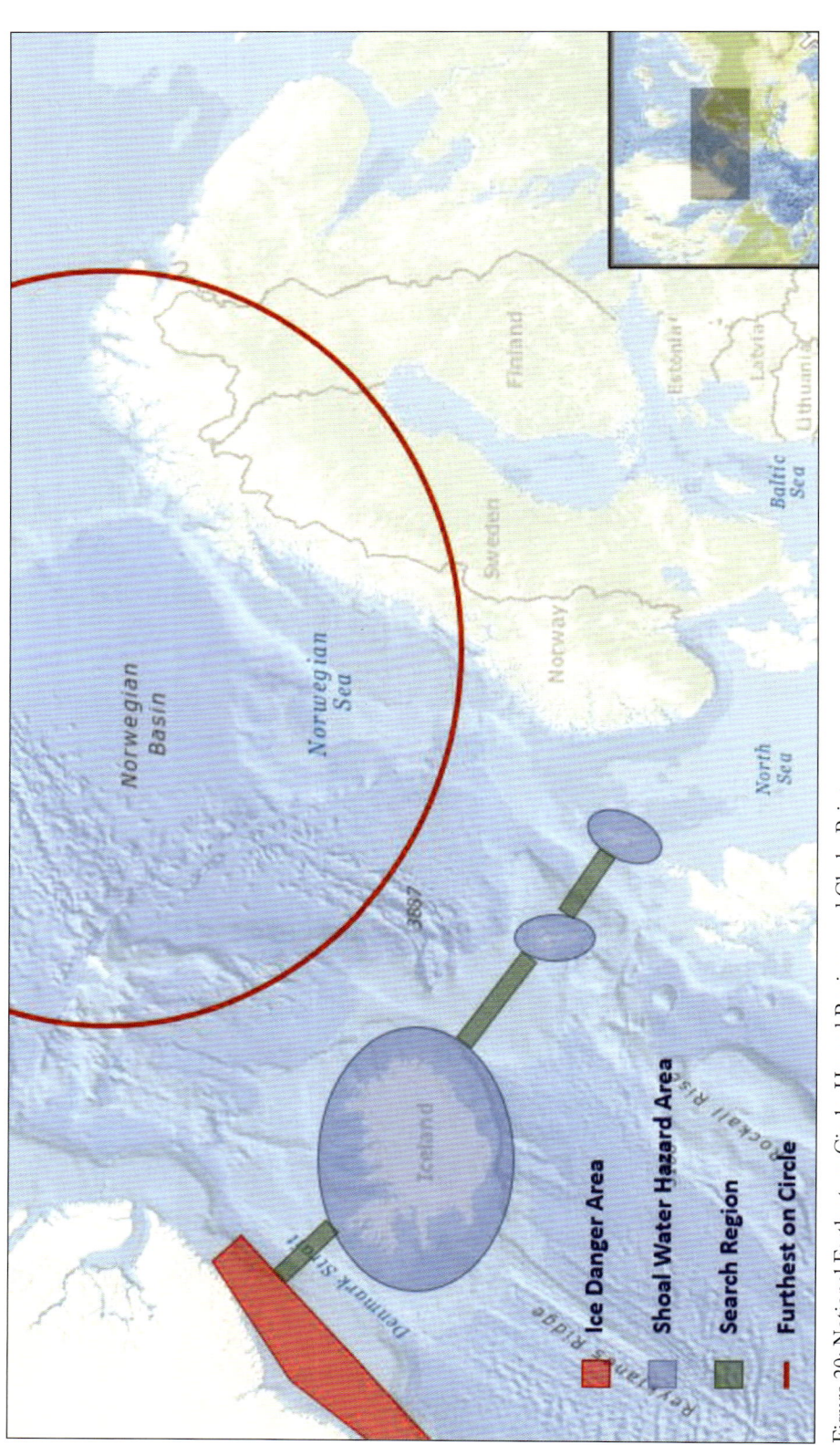

Figure 20: Notional Farthest-on Circle, Hazard Regions, and Choke Points.

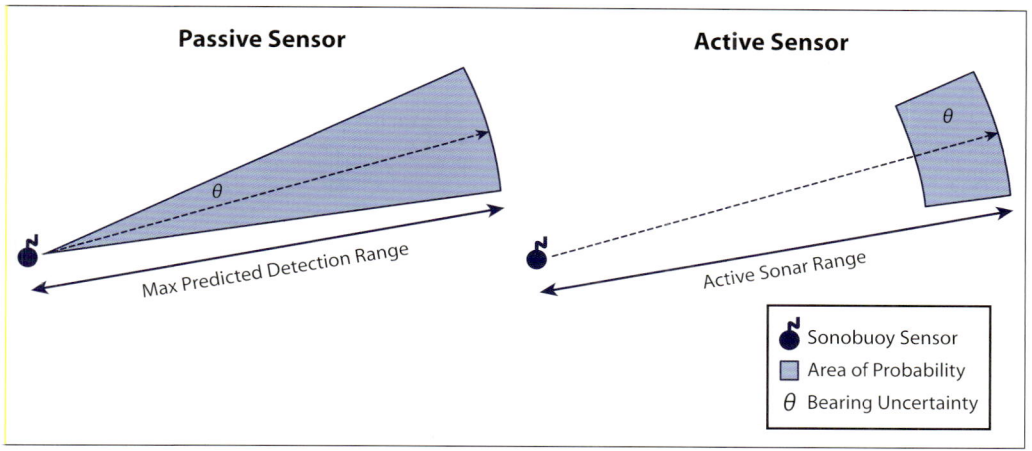

Figure 21: Passive Sensor and Active Sensor Areas of Probability.

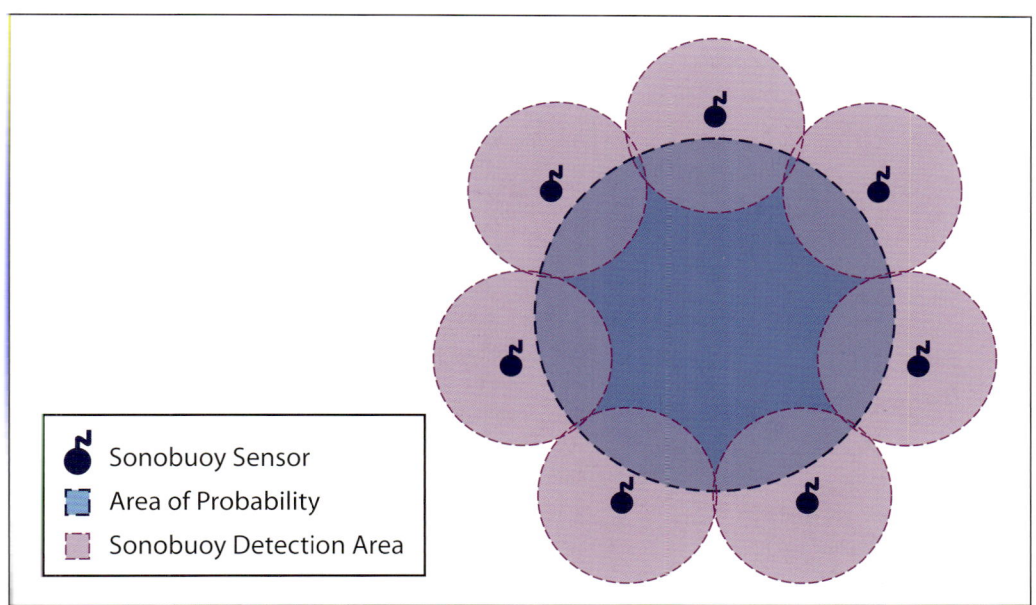

Figure 22: Circular Localization Pattern.

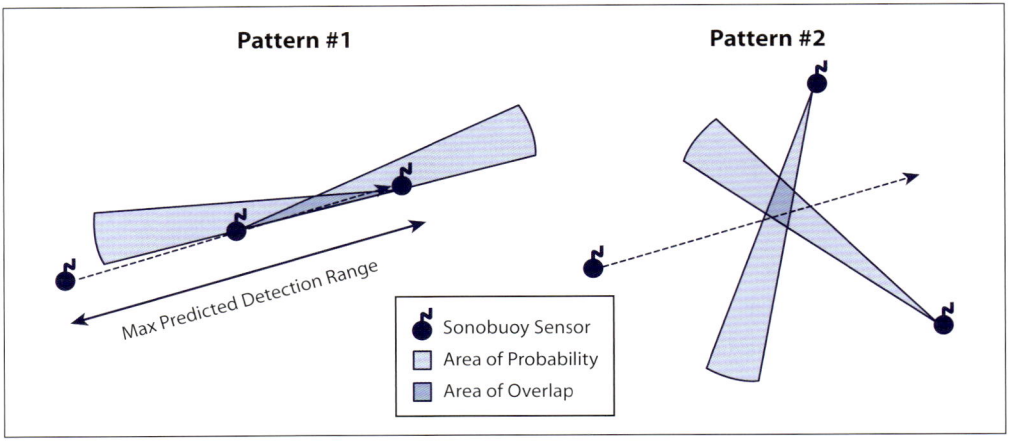

Figure 23: Localization Patterns for a Line of Bearing Detection.

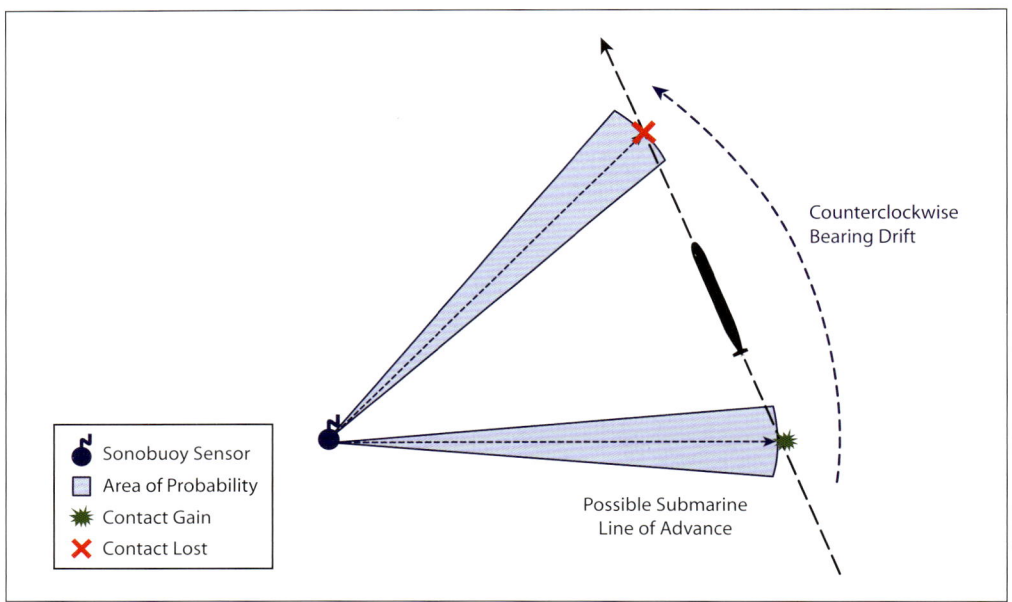

Figure 24: Gain, Counterclockwise Bearing Drift, and Lost Contact.

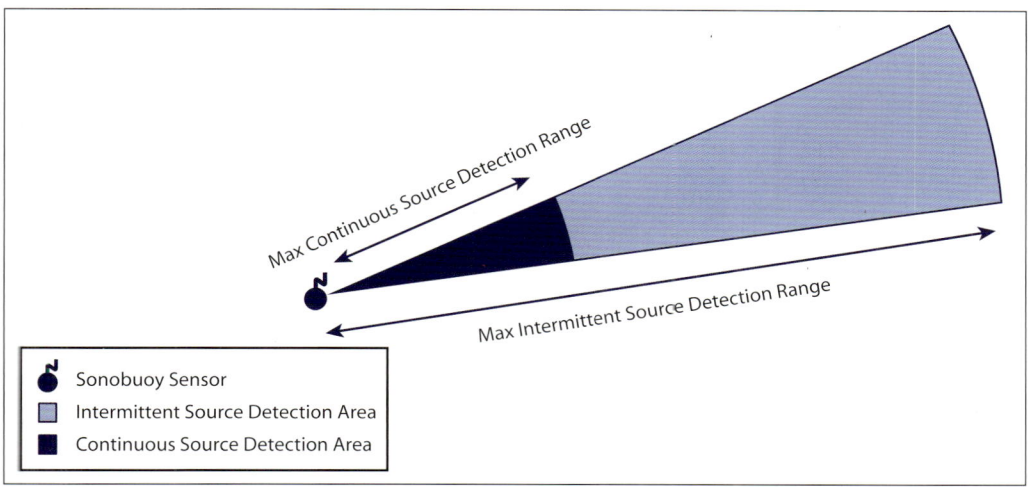

Figure 25: Area of Probability based on Intermittent and Continuous Sound Sources.

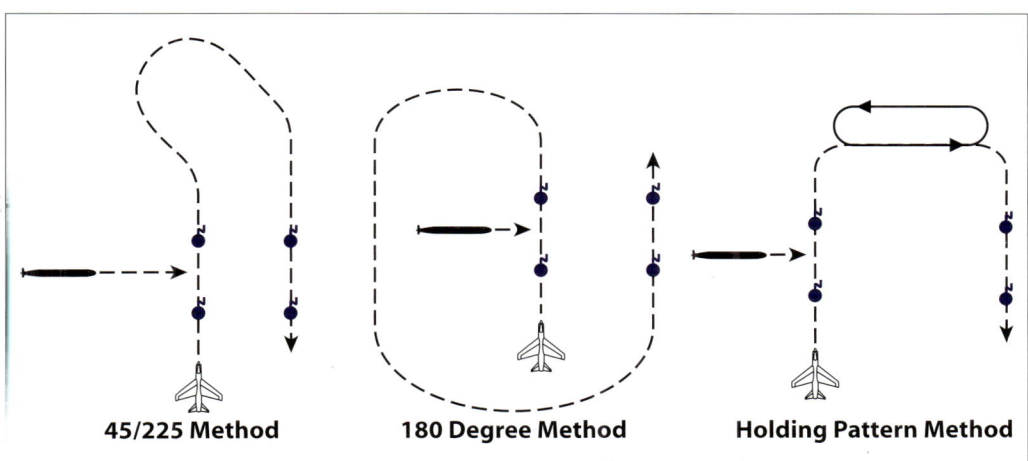

45/225 Method   **180 Degree Method**   **Holding Pattern Method**

Figure 26: Tracking Maneuver Techniques.

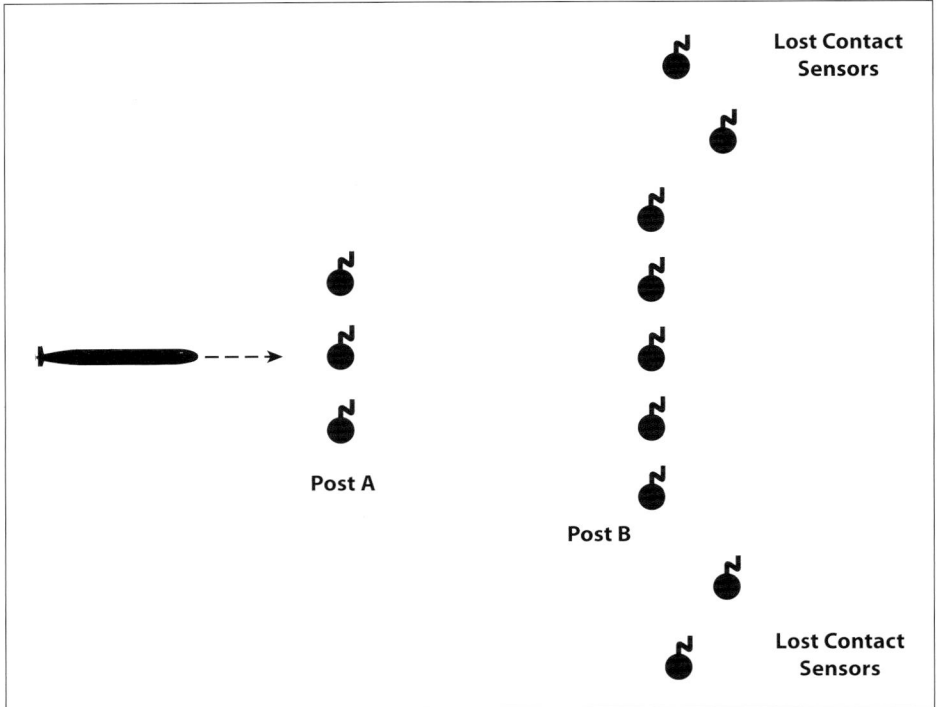

Figure 27: Post A, Post B, and Additional Sensors During Event Tracking Scenario.

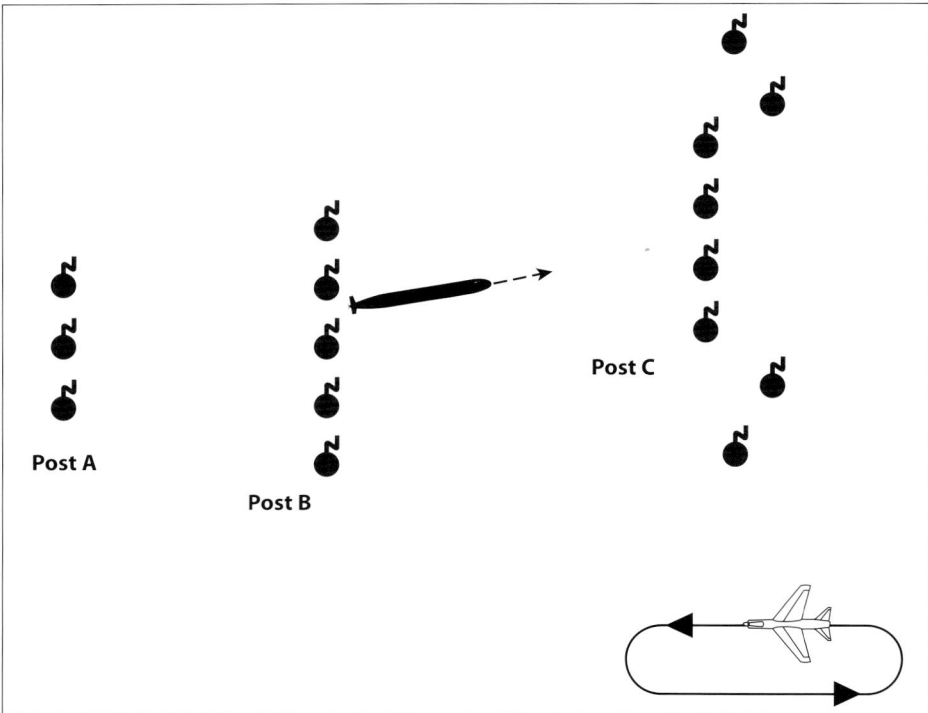

Figure 28: Post A, Post B, and Post C During Event Tracking Scenario.

In this chapter, we will discuss how to calculate TL and how to conduct ray path analysis. Both skills are critical for ASW aviators. Interpreting a TL plot and ray trace helps aviators visualize how sound is propagating in the water and where they should place sensors in order to gain contact with a target.

Transmission Loss

In Chapter 7, we discussed at length the passive and active sonar equations. As a result, we are familiar with the variables in each of them. In Chapter 8, we covered how the SSP results in sound refraction and how the ocean surface and seafloor reflect sound waves. By putting our knowledge together, we can determine how a particular sound ray will move as it radiates outwards from a source. We can also analyze how acoustic energy will lose intensity as it spreads outwards from a target and is refracted, reflected, and attenuated by the ocean environment.

Analyzing the direction in which a particular sound wave or group of sound waves move is known as ray tracing. While building ray traces, acousticians use mathematical methods that take into account the SSP and bathymetric data. By applying Snell's Law and analyzing reflections off the ocean surface and seafloor, an acoustician can break the motion of a sound ray down into many hundreds of refractions and reflections. While the calculations are trivial from a mathematical perspective, they are computationally intensive. Ray tracing can be done manually, but it is very tedious, so modern ray tracing is conducted using computer programs. **Figure 13** shows an example of a SSP and a ray trace.

Before discussing TL calculations in a tactical scenario, let us review the variables in the passive sonar equation. First, we will focus on the variables that are defined by the target submarine and the sensor in use. We can find the estimated source level (SL) of the submarine for the particular frequency of interest from an intelligence database. The Directivity Index (DI) is a set value for the hydrophone in question while the Recognition Differential (RD) is a function of the capabilities of the sensor, acoustic processor, acoustic display, and team of operators. Values for DI and RD can be found in technical reference publications.

$$SE = SL - TL - NL + DI - RD$$

Now, we can move to define the variable driven by the environment, namely ambient noise (AN). From our discussion in Chapter 8 we know that we can predict an AN value for our operating area based on historical values, expected shipping levels, and forecast wind speed. These values are either stored in databases or derived from working curves such as the Wenz curves or Knudsen Spectra. Once we have entered an AN value, we are left with Signal Excess (SE) and TL remaining.

As sound waves travel away from a target, they suffer losses due to spreading, scattering, and attenuation. Because we can predict the various processes by which sound waves lose intensity, we can calculate TL as a function of range from the sound source. Since each wave radiates from the source with a particular intensity, we can predict how the signal will lose strength as it travels. This allows us to plot how much loss has occurred as a function of range.

Simply put, SE represents the strength of a signal at a particular range when it reaches a target. SE drives how likely or unlikely an operator is to detect a signal of interest. If sound waves arrive at a sensor with high SE, the event will be presented very clearly on sonar equipment, providing the operator with an easy detection. A sound wave arriving with low SE will be more difficult for the operator to pick out from the background AN.

Before we continue, we must be very clear about the limitations of any TL calculation. A TL plot is only accurate for the conditions used to construct it. It is valid for a particular frequency, the SL of the target at that frequency, the estimated AN at that frequency of interest, the characteristics of the sensor being used, the SSP, and the bathymetric features in question. Changing any of these variables will alter the TL plot. Aviators who attempt to use TL calculations generated for a particular set of parameters at a different location or when the AN is different will often find their range predictions to be very different from actual values observed on station. The importance of using appropriate values for the different variables in a TL calculation cannot be overstated.

Figure of Merit

During the development of early sonar systems, engineers and acousticians sought to measure how well or how poorly a particular hydrophone was fulfilling its goal of detecting a particular sound source. They coined the term Figure of Merit (FOM) to describe the effectiveness of the sonar when paired with a particular SL from the target and the environmental conditions in the test area. ASW aviators have come to use FOM to gauge sensor performance. It is customary for aviators and planners to analyze the instance where the sonar is just barely detecting the target, since this defines the outer limit of the detection range.

To reflect this scenario, ASW planners set SE to zero and re-arrange the passive sonar equation. Based on the SL, RD, DI, and AN in the area, they are able to calculate a FOM for the particular scenario. By convention, FOM is assumed to provide an alerted operator with a 50 percent probability of detecting a signal.[1]

$$FOM = SL - AN - RD + DI$$

A Deep-Water TL Plot

To explore the TL plot and concept of FOM further, let us examine a common deep-water sound propagation scenario. Consider a part of the deep ocean with a moderately-sized mixed layer, thermocline, and cold abyssal water below. These conditions generate a SSP already familiar to us from our discussions of the convergence zone arrival path. The seafloor is smooth, resulting in low bottom losses, and there are sufficient depth excess and sound speed excess to support convergence zone formation. **Figure 14** shows a TL plot using a notional target submarine and senor pair and two different AN levels (i.e., resulting FOM of 69 and 72) to examine the plot of the signal strength versus range from the target and how AN impacts sensor performance.

What do we see in the TL plot when observing the lower AN value? On the Y-axis we find signal strength measured in dB. On the X-axis we find the range from the target submarine measured in kilo-yards (kyds).[2] Moving from left to right across the chart, we find the FOM = 72 index intersecting the signal strength plot at several points. Starting at the Y-axis, we find signal strength falling away until it crosses the FOM = 72 value. This distance from the target marks the longest range that the sonobuoy can detect the target via the direct path arrival mode.

The signal strength plot continues to decrease until we observe a slight increase. This marks the area where sound energy bouncing off the bottom is returning upwards towards the depth of the sonobuoy sensor. Sound energy is focused, creating a higher SPL than we would see otherwise at this range. However, the signal strength is below the FOM, meaning that our sensor will not detect enough sound arriving via bottom bounce to register a detection event for the operator.

Moving further right on the plot, we find two peaks in signal strength at a range of approximately 60 kyds and 120 kyds. These peaks mark the first and second reswept zones, where convergence zone propagation funnels sound rays together to create two highly ensonified areas. Because the SPL here climbs back above the FOM, a sonobuoy will be able to detect the signal. The range from the target where the signal strength plot crosses the FOM allows us to define the inner and outer edges of the reswept zone.

Let us imagine we could view the scenario from above and plot the parts of the ocean where we could detect sound traveling outwards from this target submarine in this particular water column. We would find a small inner zone surrounding the boat itself. In this area, a sonobuoy could detect direct path contact. Beyond this small area, we find concentric rings that correspond with the convergence zone annuli. A sonobuoy in these areas would be able to detect the submarine as well.

We must keep in mind that the signal strength plot above is only valid for the assumptions we made while calculating TL. To drive this point home, let us increase the AN at the frequency of interest while leaving all other parameters in the scenario the same. When we view the TL plot and FOM = 69 in **Figure 14**, we can see that the FOM is higher up on the Y-axis. We note that the signal strength plot crosses the new FOM at a shorter range, denoting a shorter direct path detection range. We also find that the signal strength peaks corresponding with the first and second convergence zones do not intersect the new FOM, but rather sit lower. This means that while the SSP and water column support convergence zone formation, there is insufficient SE to exploit this phenomenon.

In the previous examples, we have assumed that the acoustic signature and resulting source levels of the target boat do not change. This is not a completely realistic assumption. A submarine crew may change the engineering lineup of their propulsion plant, turning certain equipment on or off or shifting to various modes of operation. The target may operate at different speeds, which creates complications for planners, since different speeds will cause variations in the sound signature due to phenomena such as cavitation and flow noise.

For example, an aircrew might plan to detect a target SSN by the signature of its reactor coolant pumps. If the target is moving quickly, its pumps will be running at high speed and generating correspondingly high noise levels. If the submarine crew slows down and shifts the pumps to low speed, the intensity of the sound generated by this source will drop and the corresponding SL will decrease.

To better understand this concept, let us consider a target SSGN in two different mission scenarios. In the first situation, the boat is expected to be conducting a surveillance and intelligence-gathering mission. To employ its sensors, the boat will likely be at periscope depth, sailing slowly, and operating its reactor coolant pumps at low speed since very little power is required. The aviators will therefore model their source above the mixed layer and rely on tonals produced by the coolant pumps and other machinery.

In the second scenario, consider the same SSGN transiting from its homeport to a patrol area. The boat will likely be moving at medium to high speed well below the mixed layer. Since the boat is moving quickly, the aviators assume the target will likely be cavitating, will probably be producing a fair amount of flow noise due to water moving over the hull, and will be running its coolant pumps at high speed to remove large amounts of waste heat from the reactor. The aviators can therefore model TL based on the higher source levels from the high-speed pumps, flow noise over the hull, and cavitation from the screw. It should be clear that planners must make accurate predictions about the manner in which the target boat will operate if they are intent on accurately predicting signal strength and TL.

A Shallow-Water TL Plot

To examine how bathymetry affects sound propagation and TL, let us move our notional target submarine to shallow water. While the shallow ocean close to shore is often very noisy and has a different SSP compared with deep water, we will keep AN and the SSP identical to the deep water scenario we examined previously. Let us examine how simply changing the depth of the water changes TL, the predicted ray paths, and what sound arrival paths an aircrew can exploit.

When we view the plot in **Figure 15**, we find signal strength falling off rapidly. When signal strength first intersects the FOM, we define our maximum direct path detection range. Signal strength continues falling off before rising rapidly in a strong peak. This peak is caused by sound waves being refracted downwards by the negative sound speed gradient, reflecting off the seafloor and returning upwards towards a shallow sonobuoy. Because the bottom is much shallower than our deep-water example, sound waves lose much less of their intensity as they travel from the submarine to the ocean bottom and back upwards towards the sensor. This ensures that signal strength is high enough to climb back above FOM. As a result, aviators can exploit direct and bottom bounce arrival paths in this water column

Because the ocean is shallower in this scenario, there is insufficient depth to allow a positive sound speed gradient to develop. Without this gradient there is no critical depth and therefore no depth excess or sound speed excess to enable convergence zone or DSC phenomena to develop. As we move right on the plot beyond the bottom bounce zone, we find signal strength falling off asymptotically.

Let us visualize an overhead depiction of where a target could be detected relative to a sonobuoy. We would find a circle surrounding the target where aviators could place sonobuoys to generate direct path contact. Beyond this circle there is a wide ring where bottom bounce contact is possible. If we compared this bottom bounce ring to the CZ annuli we modeled in **Figure 14**, we would find them located at a much shorter range.

The Full Field Plot

Having discussed TL calculations and ray tracing at length, we can now move to combine these two useful concepts into a single tool known as the Full Field plot. The Full Field combines the ability to predict and visualize propagation paths offered by ray tracing with the ability to predict and visualize signal strength offered by plotting TL. When building a Full Field plot, acousticians simultaneously conduct a ray trace for a particular target, environment, and sensor combination. They also conduct a TL calculation to define signal strength as a function of range and location from the target.

The results are fused together in a plot that uses rays to depict how sound will propagate relative to the target and a color scale to depict the levels of signal strength at various ranges and depths. A key benefit of the Full Field plot is that a particular color can be used to depict a certain FOM. As a result, a Full Field plot can be used in-flight to solve for detection range even if the in-situ AN value is different than the predicted value used to construct the Full Field plot. This gives aviators flexibility on station. The Full Field plot is an excellent tool to quickly visualize and quantify sensor performance. An example may be found in **Figure 16**.

Directional Variation in Transmission Loss

At times, a sensor may be operating in a homogenous water column, where the SSP, AN level, and bathymetry are constant everywhere in the operating area. However, the ocean is a very variable environment. With that in mind, let us observe a notional target from several different directions in an area with changing environmental features and note how these features alter sound propagation paths and TL.

Consider a submarine operating in the deep ocean near an underwater ridge. Over the abyssal plain, the water is deep enough to support convergence zone conditions and the boat is noisy. Over the ridge, however, the seafloor rises high enough to remove the depth excess required for convergence zone propagation. Let us imagine a submarine operating west of the ridge, half a CZ range away, and two sonobuoys nearby, one over the ridge and another one CZ range west of the target.

Figure 17 shows the Full Field from each of these two scenarios. To the west of the boat, we see sound being channeled into a convergence zone, there it is picked up by our sonobuoy. To the east, we see sound waves initially refracted downwards and channeled by the duct. However, the sound waves then hit the rising seafloor of the ridge, where they are absorbed, scattered, and reflected upwards. This is a good example of how sound propagation is directionally variable. It is not enough for aviators simply to pick a random direction from a target, generate a TL plot, and then go fly a mission. They must be aware of how the environment and bathymetry will drive propagation and influence the tactical scenario.

Detection Area with Respect to a Sonobuoy

Up to this point, we have examined the case of a submarine radiating sound into the water and have analyzed TL and signal strength with respect to the submarine. We have used this information to plot areas around the submarine where detection with a sonobuoy is possible. Now let us examine the reverse scenario, where we know the location of a sensor and wish to plot areas where we expect to be able to detect a submarine relative to the sensor should a target pass nearby.

Consider two sonobuoys, one located in shallow water on the Continental Shelf and the other located over an abyssal plane that is deep but not sufficiently so as to support convergence zone contact. On the Continental Shelf, the combination of SL, SSP, and bathymetry support direct path and bottom bounce contact. In the deep water, the ocean floor is too deep to support bottom bounce but not sufficiently deep to allow convergence zones to form. This means that only direct-path contact is possible. However, because AN values are lower out in the deep water, direct-path contact is possible over longer ranges compared with the shallow water on the Continental Shelf.

When viewed from above, the detection areas around the shallow buoy appear as a circle of direct path contact and an outer ring of bottom bounce contact. In the deep water, the detection area appears as a larger circle of direct-path contact.

In reality, these detection areas are even more variable due to changes in bathymetry and AN values in the various sectors around a sonobuoy. Recall our discussion earlier on the directional variation of TL. Depending on the direction, bathymetry may allow for different detection ranges in different sectors. Additionally, AN may be higher in certain sectors and lower in others. Therefore, when modeling detection range with respect to a sensor rather than with respect to a target, it is better to analyze several radial sectors around each sensor.

To calculate appropriate detection areas, computer programs are typically used to conduct a TL calculation and ray path analysis in several sectors around each sonar array. A FOM is calculated and used to define the limits of direct path, bottom bounce, and convergence zone contact. These areas are then displayed for each sector around each sensor in a planned operating area.

As we mentioned previously, the tonals and sound sources emitted by a submarine will vary depending on the way that a submarine is operating. If a target boat shifts to a different propulsion lineup, alters its speed, or changes depth, the detection areas relative to a sonobuoy will change as well.

Using TL Calculations Operationally and Tactically

Calculating TL and detection ranges helps aviators to plan and employ sensors as effectively as possible during the different phases of a mission. These calculations also assist commanders as they devise search plans and determine operational measures of effectiveness.

During search planning, TL values are calculated at various locations in the search areas using characteristics of the target submarine, environmental forecasting databases, and bathymetric data. Detection ranges at various locations are used to calculate the probability of detection during the search. When aviators arrive on station to begin a search, they deploy sensors to measure the SSP and AN. This allows them to re-calculate TL and detection ranges and determine if their original search plan is still valid or should be adjusted. We will discuss search planning in greater detail in Chapter 12.

During localization, aviators drop additional sonobuoys to determine what sound arrival path exists and to better fix the location of the target. When aviators make an initial detection, they must "work backwards" from their sensor, determining where the submarine might be located so as to generate contact at the sensor. For example, if characteristics of the detection signal suggest bottom bounce arrival, aviators must use a Full Field analysis to determine detection range and where the boat might be relative to their sensor to result in sound waves arriving at the sensor. They can then drop new sonobuoys near the expected location of the target to gain direct-path contact and transition to tracking. We will discuss localization procedures in Chapter 13.

Aviators usually track a submarine using short-range direct-path contact. This allows them the most accurate position information so they can gather acoustic intelligence (ACINT), accurately plot position, course, and speed, or drop a weapon. The geometry of sonobuoy tracking patterns is highly influenced by detection range. Because the environment and behavior of the submarine may change during a mission, aviators must constantly monitor the environment and update their predictions of detection range. This helps them determine where the next set of sonobuoys should be dropped to maintain contact.

Last, ASW aviators make every attempt to maximize the probability that their torpedoes acquire a submarine and home to impact during an attack. TL calculations help model the performance of a LWT seeker head and forecast how likely the weapon is to acquire the target. Analyzing the environment and sound propagation helps aviators to place the torpedo at the optimum location relative to the submarine to maximize the probability of acquisition and the chance for a kill.

Notes

1. For a more in-depth discussion of each of the terms of the passive sonar equation and background on the development of FOM, readers should see Robert J. Urick, *Principles of Underwater Sound* (Westport, CT: Peninsula Press, 1983), pp. 17–30.
2. Western aviators typically discuss detection ranges in terms of yards, in keeping with naval conventions. As a result, most Western TL plots are displayed using kilo-yards to measure distance.

Chapter 10

Intelligence Gathering and Cueing

Finding a submarine in the open ocean is a very challenging task. The stealth of modern submarines means that the detection range of tactical sensors is usually quite short. As a result, ships, submarines, and aircraft need intelligence information to alert them to the general location of a target boat and to vector them into a proper search position. Without a system of long-range surveillance sensors to gather information on a target submarine and a team of intelligence analysts and operations planners to develop search strategies, an ASW operation is likely doomed to fail

Since the dawn of ASW, aviators and planners have relied on intelligence-gathering efforts to learn about their targets, inform them when a boat is underway, provide search planning support, and help direct vulnerable ships away from areas where submarines are hunting. Understanding how intelligence-gathering systems, sensors, and command and control constructs function is crucial for aviators to comprehend their role in a larger ASW effort.

In this chapter, we will begin by reviewing some basic concepts of intelligence gathering and discuss how cueing information is developed, protected, and disseminated. We will review the various sensor types used to generate indications and warnings (I&W) and provide cueing information to aircraft. Last, we will review five case studies of past ASW campaigns to learn more about how sensor systems and intelligence services have performed in the past.

Modeling Adversary Behavior

A key role of intelligence analysts is building a holistic understanding of how an adversary submarine force operates. This includes characterizing how the foreign force trains its sailors, educates its commanders, maintains its equipment, and conducts research and development of new submarine technology. Understanding the "lifecycle" of individual submarines is another critical task. This requires analysts to review the operating history of a boat from the time its keel is laid to the time it is retired and disposed of. Tracking when an individual boat is in operational status, how prepared it is for a combat patrol, when it requires extended time in the shipyard for overhaul, and when it is nearing the end of its service life helps analysts determine what type of exercises or patrols might be carried out in the near future.

Analysts also build models that characterize how submarine crews operate once they are out at sea. This practice came of age during World War II, when Allied operations analysts built sophisticated models that described every step German U-boats took from construction, to combat patrols, to shipyard overhauls, and to subsequent patrols.[1] Crews operate in several different modes while underway, depending on their objectives. We will discuss these operational models in further detail while reviewing search theory in Chapter 12.

I&W and Cueing

Before continuing, it is useful to discuss what we mean when we use the terms I&W and cueing. In a purely doctrinal manner, Western militaries and intelligence agencies use the term I&W to describe intelligence analysts warning military commanders and civilian policy makers that an adversary is more likely to carry out dangerous or aggressive behavior based on observations of shifts in the adversary's operating posture. In more colloquial terms, aviators and ASW planners use the term I&W to describe indications that a target submarine is preparing to undertake a patrol or is likely to put to sea in the near future.

Cueing, on the other hand, describes the use of long-range sensors to alert tactical forces to the presence of a submarine and to vector tactical assets to a search area to refine the location of the target. In the following sections and case studies, we will review how intelligence-gathering sensors are used for providing I&W of submarine actions and cueing to search forces.

Security of Sources and Methods

It is paramount that intelligence-gathering organizations protect the sources and methods that they use to collect information. This remains especially true in the case of ASW sensors, because of the inherent power of submarines and the sensitivity of ASW operations. Revealing the manner by which intelligence organizations or navies detect submarines and cue search forces can allow an adversary to develop countermeasures or evade the coverage area of cueing sensors.

As a result, it is neither beneficial nor desirable for aviators to have detailed knowledge of the technical capabilities of their cueing sensors. What is desirable, however, is for aviators and planners to have a basic understanding of the types of sensors used for cueing and to be able to conceptualize how the process of gathering and disseminating cueing information impacts search operations.

The security of sources and methods are protected by several means. First is the concept known as "need to know." This precaution ensures that only those personnel with duties that require them to possess a piece of information should be given access to that information. Second is the concept of compartmentalization. Compartmentalization divides information up into access areas and limits the number of people who have a complete understanding of how a particular system or organization as a whole functions. Third is the concept of obfuscating the technical capabilities of a sensor to those who do not have a need to know the true capabilities of a sensor.

For example, let us assume an ASW force has a long-range sonar system that is capable of detecting foreign submarines in a particular area of interest. The existence of this sensor will probably be divulged to ASW aviators before they begin a search mission. However, its existence will not be advertised far and wide to personnel engaged in other warfare areas. Second, planners might be given information on the regions of nearby operating areas where the sensor can search well and where the sensor performs poorly. Aviators, on the other hand, do not need to know this information, since they will merely be receiving cueing. Sharing this information with more people raises the risk it will be divulged. This is an example of compartmentalization. Last, planners must be careful not to divulge the true performance characteristics of the sonar system. To obfuscate the capabilities of the sensor, they might pass search areas to aviators in flight that are larger than the actual cueing area developed by the sensor. This is a good example of protecting the source and method of intelligence collection by obfuscating the technical characteristics of the sensor.

Timeliness and the Quality of Search Cueing

Cueing information is often derived from highly sensitive sources. As a result, the information must be handled in appropriately secure channels, ensuring that care is taken in handling the information protects the data.

However, it also means that extra time and effort are required to move the data from sensor operators to analysts, planners, and aviators.

Keep in mind that ASW cueing data is generated by submarine targets that are usually mobile. While cueing data is being analyzed, handled, and disseminated to search forces, the target submarine continues to move. Cueing data is combined with other tactically relevant information to generate an AOP where the submarine is likely located.[2] The initial size of the AOP is driven by the technical characteristics of the sensor. However, once the sensor ceases observing the target, the AOP begins to grow based on the speed at which the target can sail.

Thus, it is clear that the size of an AOP and the quality of the search planning information available to aviators are directly tied to the speed at which cueing data can be ingested, analyzed, handled, and disseminated to planners and tactical operators. Intelligence organizations and navies must strike the right balance between protecting their sensitive information and ensuring that their tactical forces have access to timely and high-quality cueing information.

Survivability of Cueing Sensors

The capabilities of long-range ASW cueing sensors make them extremely valuable to intelligence-gathering organizations. Often, these sensor systems are complex, expensive, and small in number. They may be stationary, as is the case with sensors on the seafloor, or predictable in their movements, as is the case with reconnaissance satellites. They may also be fragile or unable to withstand damage. As a result, planners must take into account the survivability of their cueing systems in wartime.

ASW cueing systems often make use of seafloor sensors or satellites. The depth and altitude of these systems makes them impervious to many threats, but both types are still vulnerable to attack by advanced nations. Planners must assume that adversaries will go to great lengths to degrade and destroy stationary sensors or those with limited mobility, such as satellites.[3] Mobility does offer some survivability to cueing systems, but to ensure safety, mobile sensor systems should also be well armed. For example, an unarmed surveillance ship towing an advanced sonar array will be more challenging to locate than a fixed hydrophone array but will still remain highly vulnerable to any armed adversary. A warship towing the same sonar system will ensure that the sensor is much more survivable compared with one towed by an unarmed ship.

Types of Cueing Sensors

Intelligence-gathering agencies take advantage of several types of sensors to learn about adversary submarines. In the next section of this chapter, we will discuss these various intelligence methodologies and some of the advantages, disadvantages, and considerations that must be taken into account when using each particular type of sensor.

Imagery Intelligence

Imagery intelligence (IMINT) makes use of EO cameras, IR cameras, multispectral imaging systems, and radars to observe submarines and their support infrastructure. While IMINT cannot locate boats underwater, it is a very useful tool to observe the day-to-day operations of an adversary force. It can provide I&W of upcoming submarine patrols and can provide useful insights into the overall health and activity of an adversary submarine force.

IMINT is collected primarily by reconnaissance satellites and secondarily by aircraft. EO cameras are used to image ports during daytime. IR cameras may be used during daytime and nighttime. They are useful for determining if ships or submarines are getting underway because they can detect the heat signature of an operating propulsion plant. Radar imagery is generally not as detailed as EO or IR imagery, but may be used during night or times of cloud coverage, making it a valuable sensor.

IMINT requires a large support infrastructure of analysts to examine the resulting photographs and interpret operationally-relevant details. This can be a time-consuming process. However, IMINT is excellent for determining if a boat is preparing for an upcoming training cruise or patrol. For example, if imagery shows a boat moving into a dry dock, it is likely that the submarine is entering a long-term maintenance period during which it will be unavailable for operations. A boat photographed while moored at a degaussing barge is probably having its magnetic signature adjusted to decrease the effectiveness of MAD equipment in preparation for a combat patrol in the upcoming months. A boat that is undergoing weapons loading is likely to depart port in the next few days for a torpedo exercise or a combat patrol. These clues help provide I&W to an ASW force and alert them to upcoming operations.

Navies are well aware of the intelligence value of IMINT. As a result, submarine forces often go to great lengths to hide their activity from the prying eyes of foreign intelligence gathering forces. Many submarine bases worldwide feature tunnels or covered docks to shelter submarines. These structures allow commanders to hide their boats from intelligence-gathering satellites and aircraft. If a boat is missing from imagery as a satellite passes overhead, analysts cannot be sure if the boat has departed for a patrol or is merely inside a tunnel undergoing maintenance. This complicates analysis and operational planning.

Signals Intelligence
SIGINT is the interception and exploitation of electronic signals emitted by a variety of equipment. Because methods of exchanging information or using radiofrequency phenomena have so many various applications, SIGINT is a very broad field with many sub-disciplines. We will focus first on the interception and exploitation of signals themselves. Later in the chapter we will discuss certain SIGINT sub-disciplines. These include communications intelligence (COMINT), namely analysis of communications transmissions to gain useful information. They also include cryptanalysis, the process by which encoded transmissions are deciphered to yield useful intelligence insights.

One of the most useful applications of SIGINT is the ability to intercept and plot the location of a transmission of interest. This can be accomplished by overlapping bearings generated by multiple intercept sites. It can also be done by utilizing more sophisticated analysis methods such as time delay of arrival that allow a signal to be geo-located. Regardless of the method of receipt and analysis, a signal broadcast by a submarine that is subsequently intercepted and plotted is a very useful cueing source for ASW operations.

The accuracy of SIGINT location data is generally a function of the technical characteristics of the receiver and the properties of the signal itself. Some intercepts may provide location data accurate enough to define a search area. Other intercepts may be of lower quality, and only useful when combined with other information to provide context or trend information. Often, SIGINT intercepts produce ellipses on the earth's surface that show the area from which a signal was likely to have originated.

Because SIGINT is a reliable and long-range sensing method, submarines often go to great lengths to limit their transmissions. A boat conducting a training cruise close to its home base is likely to transmit updates and information to commanders ashore because its location and operations are not particularly sensitive. However, a submarine crew on a combat patrol is likely to maintain radio silence or limit their communications to the bare minimum required by commanders ashore.

Aviators should keep in mind that their movements and operations can be detected and tracked using SIGINT as well. Patrol aircraft radiate through a variety of electronic means, including radio transmissions to air traffic control, radar altimeters, data links, and search radars. These signals may be exploited by land-based or space-based systems operated by foreign intelligence-gathering organizations or detected by ESM equipment onboard a target submarine itself.

Communications Intelligence

COMINT is a subset of SIGINT that focuses on the interception and analysis of communications transmissions. It is highly relevant to understanding the operations of the support forces that maintain and prepare submarines for operations. It can also provide I&W of an impending departure or arrival of a submarine at a port.

Maintaining a fleet of submarines is an extensive undertaking that requires the work of hundreds or thousands of administrators and support personnel. If intelligence-gathering organizations can tap into the radio, telephone, microwave, and cellular nets used by shipyard personnel, navy administrators, and crewmembers, then they can gain crucial insights into the operations of the adversary submarine fleet itself. Information on maintenance, training, and personnel can help analysts understand the lifecycle of individual boats and the manner in which the submarine force as a whole operates.

COMINT can also be used as an I&W tool. A boat preparing to depart port needs the assistance of a variety of support forces. These include line-handlers, tugs, harbor pilots, and security boats. COMINT intercepts can detect the radio transmissions used by these personnel and alert commanders that a submarine may be about to depart port.

Planners and aviators should keep in mind, however, that any nation sophisticated enough to operate submarines is also capable of conducting deception operations to throw off intelligence-gathering organizations collecting COMINT. For example, several individuals could easily mimic the normal communications made during a submarine departure. Alternatively, navy commanders can order the submarine and the support ships to maintain strict radio silence when the boat is about to depart on a combat patrol or sensitive mission. Lastly, should commanders have reason to suspect that their communications are tapped or monitored, they can always broadcast false or misleading information to throw off listening analysts.

Cryptanalysis

Cryptanalysis is the process by which mathematical methods are used to decode intercepted transmissions to study the information contained in those messages. Cryptanalysis is a continuous race between encryption methods on one side and decryption techniques on the other. The ability to read the contents of sensitive messages is extremely beneficial to an intelligence service. Gaining insight into the plans and intentions of an adversary can change the course of a battle or provide strategic advantages that may tilt the balance of an entire conflict.

As a result, methods of cryptanalysis are jealously-guarded secrets. The information derived from cryptanalysis is also often highly classified, and as such, is not often handled by aviators at the tactical level. More often, cryptanalysis is used to provide context to strategic and operational planning.

This use of information derived from cryptanalysis is often divorced from the tactical level of war for two reasons. First, cryptanalysis takes time. Information such as the location of a submarine might be decoded hours or days after the submarine has left that location. Information can become too "stale" to be used by search forces. Second, information derived from cryptanalysis must be tightly controlled to prevent the adversary from realizing that their communications are vulnerable. For example, should an ASW force intercept a message listing the future location of a submarine and use that data to mount an attack, the adversary submarine force may wonder how it was that an enemy ASW unit showed up at just the right location and at just the right time. Should an adversary suspect their encryption has been broken, they will move to immediately change their codes or stop transmitting all together.

Acoustic Intelligence
ACINT describes the use of acoustic signals to gain insights into the operations, tactics, construction, and capabilities of foreign submarines. For the past 70 years, sonar has been the main sensor used for ASW search, localization, and tracking operations. During the last five decades, many countries have constructed hydrophone arrays on the seafloor or deployed mobile sonar arrays towed by surface ships.[4] These systems have proven very useful for long-range cueing of tactical ASW forces.

Underwater surveillance arrays are built to perform in a certain part of the ocean and to utilize a particular sound arrival path. During both World Wars, these systems consisted primarily of short-range sensors to detect hostile submarines approaching friendly harbors. During the 1950s, NATO forces began deploying long-range hydrophones that utilized the DSC. These systems could detect Soviet submarines at very long ranges and vector forward-based ships, submarines, and aircraft into contact with target boats. During the 1980s, several nations deployed tower sonar arrays onboard auxiliary vessels. These allowed the ships to reposition the arrays and monitor areas outside the field of view of fixed seafloor sensors.

Seafloor sensors remain relatively invulnerable to conventional forces due to their depth. However, they are vulnerable to specialized deep-sea units such as submersibles and remotely-operated vehicles. These sensors are usually connected by fiber-optic cables to processing facilities ashore. This means seafloor arrays can be disabled by attacking the cables that lead to shore or attack the shore-side cable landings. As such, fixed sonar systems are vulnerable to attack in the prelude to a conflict or during wartime.

Mobile surveillance systems are quite useful due to their performance and ability to reposition. However, they are often deployed aboard unarmed auxiliary vessels rather than onboard warships. Towed arrays are often large and unwieldy, and frequently require the surveillance ship to operate at slow speeds and steam in a straight line to avoid creating excessive flow noise or curving the sonar array. The lack of mobility and armament aboard surveillance vessels makes these ships extremely vulnerable in wartime. Therefore, they serve primarily as a peacetime force to help gather intelligence and get submarines and aircraft into contact before the first shots are fired.

Datum Cueing

Datum cueing refers to using significant events that reveal the presence of a submarine. Such an event is known as a "datum," and can be used to define a search area. For example, a visual sighting of a periscope or mast can reveal the presence of a boat. Missiles rising from the water in a location where no ships are nearby show the presence of a submarine. An attack on a warship or merchant ship shows that a submarine was present within torpedo range of that location. This is known as a "flaming datum," and provides both cueing information and motivation for the aircrew to exact a measure of revenge.

Case Studies

Case 1: Use of radio direction finding by British ASW forces in World War I

During World War I, German U-boats engaged in a fierce anti-shipping campaign in the North Atlantic and the Western Approaches to the British Isles. In order to keep in contact with commander ashore, U-boats frequently transmitted messages using high-frequency radios. Soon after the beginning of the conflict, the British established a network of HF/DF intercept sites along the coast to track the transmissions of German warships.

HF/DF intercepts allowed the British to keep track of the numbers and general locations of U-boats at sea. Even if analysts could not read the messages that were transmitted, they could identify areas of high and low submarine activity. When loss rates of merchant shipping became unsustainable in 1917, the Royal Navy reluctantly adopted a convoy system. At this point, the British SIGINT efforts began to truly bear fruit.

Each convoy was controlled by a senior naval officer aboard a command ship. Equipped with a high-frequency radio, the convoy commodore could receive intelligence updates about the disposition of German U-boats. This allowed the entire convoy to maneuver away from areas of high U-boat activity. Given the short detection range and low sweep rates of U-boats, the combination of concentrating many ships into one convoy and the action of maneuvering that group of ships away from U-boats led to a drastic drop in the detection rates and sinking rates achieved by the German submariners.

British intelligence gathering and operational planning was not entirely successful during World War I, however. Royal Navy SIGINT efforts were run out of an independent and highly classified office in the Admiralty. As a result, the SIGINT office was divorced from the day-to-day actions of the operational planners. While cryptanalysts would occasionally pass notes and intelligence updates to the operational planners, this system was neither standardized nor efficient. Much of the useful information gleaned from intercepted and decoded German messages was filed away and unused or provided to operational planners who often made little or no use of it.[5] After the war, the Royal Navy was diligent about making use of these lessons learned. They resolved that intelligence derived from SIGINT and cryptanalysis needed to be tightly integrated into the ASW and convoy planning process. This set the stage for a highly effective operational intelligence plot and planning system that was utilized by the Allies in the North Atlantic two decades later during World War II.

Case 2: Use of cryptanalysis and tactical employment of SIGINT by Allied ASW forces in World War II

The work of Allied intelligence analysts and operational planners during the Battle of the Atlantic offers a very valuable case study in the utilization and handling of information gained through code breaking. While popular history emphasizes the success of British code breakers at Bletchley Park, the reality is that Allied and Axis communications specialists and cryptanalysts battled back and forth during the conflict.

During various parts of the war, one side or the other was capable of reading significant portions of the opposing navy's message traffic. Additionally, the sensitivity of the intelligence derived from British decrypts of Kriegsmarine message traffic was such that the Allies were forced to make tough choices about where, when, and how to use this precious information.

At the outset of the conflict, the Royal Navy immediately instituted a convoy system and worked to integrate HF/DF direction finding plots with operational planning. The Nazi U-boat fleet still utilized an extensive system of ship-to-shore communications. Rather than maintaining radio silence in an effort to avoid detection, German doctrine called for a submarine that made contact with a convoy to broadcast its location to headquarters and to other boats nearby. Rather than attacking immediately, this first boat in contact would stalk the convoy and wait until other submarines arrived before beginning the attack.

These extensive communications led to frequent detections of deployed U-boats by British SIGINT sites. While the Royal Navy was unable to decode the encrypted messages early in the war, they were able to fix the location of U-boats and plot the position and operating pattern of individual submarines. This allowed analysts to keep accurate and timely track of where each boat in the enemy order of battle was and where in its lifecycle of refit, training, deployment, patrol, and return to base it was. Planners were able to identify areas of high U-boat activity, route convoys around these areas, and assign additional air patrols when a particular convoy was threated.[6]

The Nazis rapidly broke the Royal Navy convoy order cipher, allowing them to read this message traffic at will. This allowed them to predict the movement of Allied convoys and maneuver their boats in front of the vulnerable merchant ships. Despite their code breaking achievement, the Germans were confident that their own communications were secure and were likely to remain so, given the sophistication of their own cipher systems and their disciplined internal security procedures.

In 1941, however, the British scored a major coup when they cracked the German naval encryption system and began reading Kriegsmarine message traffic on a regular and timely basis. While the Allies would have to suffer from sporadic periods of not being able to decrypt German message traffic due to changes in codes, this provided them with exceptional insights into the plans and intentions of the U-boat commanders.

The British enacted a tightly-controlled set of handling procedures for Nazi decrypts known by the codename ULTRA. The information derived from the decrypts was of such high quality and so timely that the Allies were confronted with a difficult dilemma. In some cases, ULTRA intercepts were sufficient to provide operational or even tactical targeting information, allowing ASW forces to predict the location of and hunt down individual boats. However, any significant uptick in U-boat sinkings would immediately alert the Kriegsmarine high command. The Nazis would look for a factor that could explain their vulnerability and would likely focus immediately on their communications systems. Any suspicion of communications security vulnerabilities would drive the Germans to change their codes or their cipher system entirely, imperiling the precious supply of information.

As a result, from 1941 to 1943, ULTRA information was used primarily to provide insights into U-boat operational intentions. It was used passively to plot the position of U-boat patrol areas and route convoys along the safest routes. In certain limited cases, ULTRA intercepts were used to hunt down individual U-boat tankers, as these boats had an outsized contribution to driving up sinkings of Allied ships.[7] The most dangerous time for a U-boats on patrol was generally its transit through the Bay of Biscay to ports in occupied France. While in the bay, submarines had to run a gauntlet of Allied ASW aircraft and their risk of being attacked or sunk was much higher. By refueling and resupplying while underway, tanker U-boats allowed Nazi submarines to remain at sea far longer, sinking many more ships than they would otherwise. Their threat was deemed sufficient to risk the security of ULTRA intercepts. As the Battle of the Atlantic tilted increasingly in favor of the Allies, American planners were willing to risk revealing their cryptanalysis capabilities by opening the information to increased chance of disclosure.

An ASW task force centered on the escort carrier USS *Bogue* (CVE-9) was provided with cueing information purportedly gained from HF/DF intercepts. In reality, the information was derived from ULTRA intercepts and decrypts. This allowed the *Bogue* and her air wing to hunt down and sink several U-boats in quick succession.

Normally, the locations of U-boats derived from ULTRA information were "sanitized" so as to appear like they were derived from normal HF/DF intercepts. This was done by providing an area of location uncertainty around the position of the U-boat similar to that which might be produced by a HF/DF intercept. In one instance, however, several American operations analysts working with their British counterparts were provided cueing information derived from ULTRA intercepts that had not been properly sanitized. The Americans quickly realized that the position area they were provided was far too precise to have been derived from HF/DF fixes. When they confronted their British counterparts about the issue, the British were forced to admit that they had a special intelligence source that could provide such location information, thereby inadvertently revealing the existence of the cryptanalysis capability.

Case 3: Use of tactical SIGINT and the effect of timeliness on search results
As we mentioned earlier, the amount of time between a SIGINT site intercepting a submarine transmission and a ship, submarine, or aircraft beginning to search has an outsized impact on how effective the search will be and how much effort must be allocated in order to find the boat. By 1941, HF/DF equipment had advanced to the point where it was small enough and reliable enough to be carried by Allied escort warships patrolling the convoy lanes in the North Atlantic. The introduction of this equipment would have a drastic effect on the lethality and effectiveness of Allied ASW forces.[8]

During the early years of the war, HF/DF sites could plot U-boat transmissions and forward this information to operations planners. However, it would take many hours to handle the information, plot the location of a U-boat, and then dispatch a ship or aircraft to investigate. HF/DF intercepts were "time late," and therefore most useful for conducting operations analysis rather than cueing search forces to respond and hunt down an individual boat.

The introduction of HF/DF equipment on warships changed this relationship completely. An escort ship could receive a transmission, plot the location of a boat nearby, and in only a few minutes have a ship or aircraft heading down the bearing towards the U-boat to investigate. Because Kriegsmarine doctrine called for one U-boat to remain in contact passing position reports on the convoy until several boats could form a wolfpack, these tactics allowed Allied warships and aircraft to frequently catch U-boats on the surface and deliver an attack.

HF/DF equipment aboard ships radically reduced the amount of time needed to get a search asset into the AOP. These SIGINT receivers were also much closer to the transmitting boat, which meant that the AOPs they developed were orders of magnitude smaller than the large search areas generated by land-based SIGINT sites operating hundreds of miles from the transmitting boat. The combination of small AOP size, timely dissemination of information to tactical forces, and the fact that ASW forces were only a few dozen miles from the target meant that tactical commanders were able to turn SIGINT intercepts into actionable tactical information rather than information for operational planning.

Case 4: Use of acoustic cueing methods by NATO ASW forces during the Cold War
During the Cold War, NATO forces made use of a very effective network of hydrophone arrays and sonar systems to provide long-range cueing information of the location of Soviet submarines. These systems served for

decades, and as Soviet submarines became quieter, the systems evolved to keep pace with the threat. Because of their capability, the information produced by these sonar systems was highly classified and tightly compartmented, so much so that many of the tactical forces that relied on them for cueing were unaware of their existence or capabilities.

When the Cold War ended, lack of awareness of these systems and their importance led to drastic funding cuts that caused many arrays to be retired and severely damaged the effectiveness of the network. The story of the NATO undersea surveillance program is therefore a very useful case study in how sensor systems develop in response to a changing threat and what is the ideal balance between protecting sources and methods and ensuring cueing sensors are timely, relevant, and well known enough to be prioritized in funding decisions.

In the years leading up to World War II, acousticians discovered the existence of the DSC and the very long-range sound propagation paths made possible by this phenomenon. The rise of the Soviet submarine fleet in the early 1950s forced NATO to prioritize ASW operations. Wary of the peril that Great Britain faced during the Battle of the Atlantic, NATO planners feared that any conflict with the Warsaw Pact in Western Europe would also lead to a third anti-submarine battle in the convoy lanes of the North Atlantic, as Soviet submarines fought to cut the lifelines of men and material from North America to the Continent.

As a result, naval leaders in NATO developed a plan to interdict Soviet boats as they departed their ports in the Kola Peninsula and steamed south to enter the North Atlantic. A series of forward-deployed ships, submarines, and aircraft were tasked to patrol choke points and prevent Soviet boats from reaching the convoy lanes. These warships and aircraft needed support from intelligence-gathering sensors to alert them to approaching submarines and guide them to their targets.[9]

Research had shown that hydrophones positioned to exploit the DSC could detect submarines at very long ranges, in some cases across entire ocean basins. As a result, the U.S. Navy in cooperation with several allied countries installed a series of undersea hydrophone arrays connected to land-based processing facilities. Known as SOSUS, the network provided effective cueing for forward-deployed SSNs and patrol aircraft.

During the 1960s and 1970s, NATO enjoyed a major advantage over the Soviet submarine force. The very high noise levels of Soviet boats allowed SOSUS to provide excellent long-range detection and cueing. However, by the late 1970s the Soviets had become aware of how vulnerable their boats were to detection by SOSUS and subsequent prosecution by NATO ASW forces. Their ensuing advances in noise quieting led to SOSUS becoming much less effective against the newer Soviet boats putting to sea.[10]

The U.S. Navy sought to remedy this deficiency in a variety of ways. One was transitioning from seabed sensors that relied on the long-range DSC to new sensors that exploited short-range RAP. Instead of installing small numbers of large hydrophone arrays, the Americans concentrated on building networks of smaller individual sensors spread over a large area and linked together.[11] The U.S. Navy also developed high-performance towed sonar arrays that were deployed on a series of auxiliary oceanographic research vessels. This system, known as SURTASS, was mobile, and therefore able to reposition to the best place to provide cueing against a target submarine.

While the SOSUS system was very effective, it had important limits that degraded its utility as an operational and tactical cueing system. First, SOSUS information was highly classified, and subject to extensive handling restrictions. This increased the latency with which detection of a foreign submarine by SOSUS could be provided to planners and eventually to aircrews. Next, NATO leaders ensured that the knowledge of the existence of SOSUS was highly restricted. As a result, a large percentage of NATO's ASW forces were unaware of the system and how truly critical it was to providing search cueing and operational support.

When it came time to make tough decisions about what systems should be cut and what should be preserved in wake to the collapse of the Soviet Union, SOSUS and its successor, the Integrated Undersea Surveillance System (IUSS) found that it did not have enough allies and supporters to fight budgetary battles and keep the arrays and processing facilities online. This led to a large number of the arrays being permanently deactivated or put on standby.[12]

Reflection by the IUSS community after the budget cuts and array shutdowns pointed to two primary factors that hampered efforts to secure funding in the post-Cold War environment. First, security restrictions put in place to protect IUSS were so severe that too few warfighters, leaders, and politicians were aware of just how critical undersea sensors were to provide effective ASW cueing. Second, what limited information was made available to tactical forces was very difficult to use. Cueing information broadcast to ships and aircraft was tabulated in messages that used unusual formats and extensive abbreviations. In the heat of action, planners and aviators simply did not have enough time to decode the confusing messages and utilize some of the cueing information.[13]

The NATO experience with SOSUS and later IUSS is a good example of a sensor system adapting to new operational realities to remain effective.

It is also a cautionary tale of how well-intentioned security restrictions can hamper the very function that sensor was built to fulfill and threaten the long-term viability of such a system. Program managers must find ways to strike the right balance between security and keeping a system so obscure that it does not cultivate a cadre of supporters and advocates who are willing to fight for continued funding and institutional support.

Case 5: Use of COMINT by NATO forces during the Cold War

During the Cold War, NATO utilized a variety of sources to exploit COMINT collections opportunities. Intelligence sharing agreements allowed NATO forces to take advantage of communications intercept sites on land in Norway and information gathered by Norwegian auxiliary intelligence gathering ships (AGIs).[14] These intercept sites could detect the communications of Soviet forces as they got underway from their bases on the Kola Peninsula. This provided additional I&W and cueing information to help operational planners vector submarines and patrol aircraft towards Soviet boats as they sailed in the Norwegian Sea.

The U.S. Navy also went to great lengths to exploit the communications of Soviet naval commanders and support personnel. The Americans allegedly utilized specially-configured submarines and diving units to place tapping pods on telephone lines in the Barents Sea and in the Sea of Okhotsk. These lines ran from naval headquarters to smaller bases and shipyards. Because the cables were in remote areas, the Soviets considered them secure, and the Americans were able to record both unencrypted and encrypted conversations. Analysts were able to listen to Soviet Navy personnel discussing submarine operations, maintenance, and other sensitive matters in frank and candid ways. This insight into the day-to-day operations of a submarine fleet and strategic nuclear force was priceless. This information was not only extremely useful in building a model of how the Soviet submarine force trained, equipped their boats, and prepared for deployments but also gave insights into how the Soviets maintained their force in peacetime and planned to employ it in wartime.[15]

Notes

1. Allied analysts studying the operations of German U-boats based in occupied France developed "Circulation Models" that described the various steps submarines took during their lifecycle and while on patrol. For more details on these models, see Brian McCue, "U-Boats in the Bay of Biscay: An Essay in Operations Analysis" (Washington, DC: National Defense University Press, 1990).
2. During the Cold War, NATO ASW forces used the term Submarine Probability Area (SPA) to denote search areas derived from SOSUS hydrophone cueing. Owen R. Cote,

Jr., *The Third Battle: Innovation in the U.S. Navy's Silent Cold War Struggle with Soviet Submarines* (Newport, RI: Naval War College Press, 2003), pp. 25–6.

3. Russia and China both maintain anti-satellite systems capable of targeting reconnaissance satellites in low-earth orbit. Russia operates a fleet of deep-submergence submarines, submersibles, and remotely-operated vehicles capable of locating and manipulating underwater infrastructure such as cables or sonar arrays. For details on the Russian Main Directorate of Deep Sea Research, see David E. Singer and Eric Schmitt, "Russian Ships Near Data Cables are too Close for U.S. Comfort," *New York Times* (25 October 2015). See also Steve Weintz, "The Ultimate Hybrid War Strategy: Attack Deep-Sea Fiber-Optic Cables," *The National Interest* (17 September 2015), http://nationalinterest.org/feature/the-ultimate-hybrid-war-strategy-attack-deep-sea-fiber-optic-13860 (accessed 8 July 2017).

4. The history and operational use of the American SOSUS system is well documented. Japan has also constructed an extensive and effective network of sonar arrays and oceanic survey vessels to provide acoustic cueing. For details of the JMSDF ocean surveillance system, readers should reference Desmond Ball and Richard Tanter, *The Tools of Owatatsumi: Japan's Ocean Surveillance and Coastal Defense Capabilities* (Canberra: ANU Press, 2015).

5. Patrick Beesley, *Very Special Intelligence: The Story of the Admiralty's Operational Intelligence Centre 1939-1945* (London: Greenhill Books, 1977), pp. 5–7.

6. Bryan Clark and John Stillion, "What it Takes to Win: Succeeding in 21ˢᵗ Century Battle Network Competitions" (Washington, D.C.: Center for Strategic and Budgetary Analysis, 2015), pp. 22–3.

7. Jeffrey G. Barlow, "The Navy's Escort Carrier Offensive," *Naval History Magazine*, Vol. 27, No. 6 (December 2013).

8. Clark and Stillion, "What it Takes to Win," pp. 9–10.

9. Cote, *The Third Battle*, p. 17.

10. For a discussion on the reduced effectiveness of SOSUS in the face of Soviet third-generation submarines, see Clark and Stillion, "What it Takes to Win," pp. 40–2.

11. Cote, *The Third Battle*, p. 78.

12. Cdr. Dawn M. Maskell, USN (ret.), "The Navy's Best-Kept Secret: Is IUSS Becoming a Lost Art" (Newport, RI: Naval War College, 12 April 2001), pp. 25–9.

13. Ibid.

14. During the Cold War and to the present day, Norway has operated a series of intelligence-gathering ships, each of which has been named *Marjata*. For details of past ships and the current iteration, FS *Marjata*, see Tyler Rogoway, "The Russian Military Despises this Strange Wedge Shaped Spy Ship," *Foxtrot Alpha* (28 October 2014), https://foxtrotalpha.jalopnik.com/the-russian-military-despises-this-strange-wedge-shaped-1648132968 (Accessed 31 October 2014). Also, see Thomas Nilsen "Spy Ship Changes Name and Continues Intelligence Mission," *The Independent Barents Observer* (30 March 2016), https://thebarentsobserver.com/en/security/2016/03/spy-ship-changes-name-and-continue-intelligence-mission (February 16, 2018).

    For details on Norwegian ELINT activities, see Norman Polmar and Edward Whitman, *Hunters and Killers Vol. 2: Anti-Submarine Warfare from 1943* (Annapolis, MD: U.S. Naval Institute Press, 2016), p. 156.

15. An in-depth and extensive discussion of the U.S. Navy submarine cable tapping program can be found in Sherry Sontag, Christopher Drew, and Annette Lawrence Drew, *Blind Man's Bluff: The Untold Story of American Submarine Espionage* (New York: HarperCollins, 1998), pp. 227–82.

Chapter 11

Crew Resource Management

ASW at both the operational and tactical levels of war is a team sport. A single submarine prosecution involves dozens, sometimes hundreds, of people working together to analyze data, exchange information, and carry out tasks. Aviators use the term Crew Resource Management (CRM) to describe the process by which members of a flight crew call on and balance the various resources inside and outside the aircraft to safely and effectively accomplish a mission. In this chapter, we will examine in detail the various CRM techniques that aviators can use to maximize their performance.

While accomplishing various parts of the ASW kill chain, aviators participate in a decision-making cycle, well represented by Boyd's Observe, Orient, Decide, and Act (OODA) loop. An aircrew must observe their sensors and the environment to gather data. They must analyze this data to produce relevant information and orient themselves to the tactical scenario at hand. They then must decide on an appropriate course of action, carry the action out, and observe the results, effectively beginning the loop again.

In this discussion, we must make the distinction between the terms data and information. Data are the raw, unorganized facts or measurements gathered by planners or operators. Once these data are analyzed, they can then be structured in a logical manner to yield operationally or tactically relevant information. Information can then be applied to solve the ASW problem at hand.

In each step of the ASW decision-making process, information must be assembled or exchanged, both by individuals and by groups of operators. Effective management of data and the process used to generate tactically usable information is the critical task placed before an aircrew. Communication is the most powerful tool that planners and aviators have to optimize information management.

Challenges to Information Management

The nature of the submarine poses significant information management challenges for ASW practitioners. Cueing data to support search operations is sporadic, time-late, and often incomplete. ASW planners and aircrews are frequently separated by hundreds of miles and therefore unable to work side by side. Communications with aircraft in flight are often limited by bandwidth restrictions and sometimes may not be available at all.

The environment aboard an ASW aircraft is often no better for information management. For much of the history of airborne ASW, crews worked onboard aircraft that were cramped, cold, noisy, and unpressurized. Even today, with climate control and pressurization, aircrews are subject to long missions, shaken by low-altitude turbulence, and suffer from motion sickness from being forced to work in a windowless fuselage as the aircraft pitches, rolls, and turns. Crews may only have an intercommunications system (ICS) to exchange information. An aircraft may bring speed, mobility, and endurance to the ASW fight, but it is a platform that makes information management and communication challenging.

While each type of patrol aircraft is unique, we can generally categorize them based on the physical layout of operator consoles and whether the tactical workstations onboard are fixed in their function or able to carry out multiple functions. Each category of aircraft will bring its own nuances and challenge an aircrew to focus on different actions as they work to optimize their tactical effectiveness.

Fixed Function versus Reconfigurable Tactical Systems

Tactical systems configuration onboard ASW aircraft can be broken down into two broad categories: fixed function and reconfigurable function. Fixed function systems are typical of legacy ASW platforms. A particular sensor is tied to a console where an operator carries out a set of tasks. The aviator sitting at a particular console can manipulate and gather data from that sensor, but generally cannot interact with sensors control by other consoles. A reconfigurable tactical system integrates the various sensors onboard the aircraft into a central tactical system. Each sensor can be manipulated and controlled by various operator consoles. A truly reconfigurable system allows any operator at any console to accomplish whatever data gathering or information management task they choose.

Variations of each type of system exist. Some ASW aircraft allow operators at one console to perform limited tasks from other consoles. Other aircraft feature fixed function consoles but allow operators to exchange information with other operators through a central tactical system. For the sake of simplicity, we will limit our discussion of fixed and reconfigurable function systems by considering whether multiple sensors can be manipulated from multiple different consoles. If sensors can be manipulated in various manners from various consoles we will call this action "load sharing" and consider such a system reconfigurable. If tactical tasks cannot be divided in different configurations and must be accomplished by dedicated consoles, we will consider such a system fixed in nature.

Fixed function tactical systems are constrained by the inability of crewmembers to alter their responsibilities and divide the workload of the aircrew as effectively as possible. Data gathered by a certain sensor is only available to certain operators, and therefore fewer eyes can analyze it. Conclusions drawn from examining sensor data cannot be checked or examined by other operators. The crew is not able to collaborate in the optimal manner.

In fixed function systems, great pressure is placed on individual operators. These aviators are forced to analyze data on their own, generate tactically useful information, and then distribute this information to the crew. Often, their only avenue for transmitting information is by using verbal communications over the ICS. Clear, concise, and unambiguous communications are critical in an aircraft equipped with a fixed function tactical system. Improving the individual performance of each operator and their ability to communicate information are the two primary methods aviators flying platforms with fixed function tactical systems can use to improve their effectiveness.

Reconfigurable function tactical systems allow an aircrew to distribute the tasks required to gather data, synthesize useful information, and employ it as efficiently as possible across all operators onboard the aircraft. An aircrew should attempt to balance duty assignments so that task loading is evenly distributed between all aviators. Tasks should fit naturally with the skills of each crewmember.

Aviators using reconfigurable function tactical systems need to divide their tasks in a way that challenges but does not overwhelm individual operators. They must resist the temptation to assign all critical tactical duties to one or two experienced operators, despite the fact these aviators can accomplish the task. More experienced crewmembers will naturally carry a heavier load than their less experienced juniors. But a junior crewmember that is tasked with too small a load will tend to unconsciously shut down and defer to the more senior crewmember. Load sharing not only spreads duties more efficiently, but also allows junior personnel to gain experience and confidence more rapidly by constantly challenging them.

Tactical tasks in a reconfigurable function tactical system need not be wedded to a particular competency area. Load sharing by its nature requires operators to have fluency in all areas of the air ASW mission and kill chain. Each operator should have a deep competence in their own sensor as well as familiarity with the capabilities of the other sensors and weapons onboard the aircraft. In this way, they can understand how operating the aircraft in a particular manner will affect not only their sensors, but also the sensors controlled by the other operators.

Aircraft Physical Operator Configuration

ASW aircraft can be further categorized based on the physical layout of their operator consoles. In some aircraft, operator consoles are spread throughout the aircraft. In other aircraft, the majority of, or all, sensor operators are located in close physical proximity to one another. Each category tends to influence how information is managed. Different optimization strategies can be applied to each type of configuration.

An enormous amount of communication between humans is non-verbal, putting aircrews with physically separated consoles at a distinct disadvantage. Some physically separated consoles have the ability to exchange tactical information in order to improve aircrew effectiveness, but generally, operators who are isolated are forced to use their voices to exchange information.

In an aircraft with physically separated consoles, a large burden is placed on each operator. Much like aviators flying with fixed function systems, they must analyze data individually, draw conclusions on their own, and communicate information verbally. They must use communications that are clear, concise, and unambiguous. The likelihood for error and miscommunication is high in this type of physical arrangement.

When operators are seated in close proximity, aircrews can take advantage of powerful synergy. Aviators can see, and in some cases touch, their fellow operators. They can take advantage of non-verbal cues and body language to pass information. With just a glance, a tactical coordinator can tell if her primary acoustic operator is task saturated and flustered or is calm and in control of the situation. A senior operator may be able to lean over and observe the sensor station next to them, helping a junior crewmember to analyze some ambiguous data. An aircrew organized in close physical proximity is much more powerful. They can complete the ASW decision-making process much more quickly and effectively with less risk of error or miscommunication.

The Tube and the Cockpit: Bridging the Gap

Even in aircraft with sensor operators seated in close proximity, there remain physical barriers to communication. The pilots flying the aircraft are posted at the forward end of the fuselage, in the cockpit. The technicians responsible for troubleshooting equipment or launching ordnance are typically located in the aft portion of the fuselage, far from the remainder of the crew. This physical separation results in significant communications barriers remaining even in aircraft with the finest configuration of sensor operator consoles.

While it may seem counter-intuitive, sensor operators sitting together often degrade effective information flow between them and the pilots. In a physically separated console arrangement, sensor operators are forced to make voice reports on ICS or to forward information to the central tactical system. In a co-located layout, operators will often chat amongst themselves, carrying on a conversation off of the ICS. These "side bar" conversations leave pilots and technicians out of the loop.

Imagine a sensor operating pointing to information on his console and saying "look at this contact here," on ICS to his colleague sitting next to him. The remark will be entirely clear to the operator sitting at the next console but will mean nothing to the pilots hearing it from the cockpit without context. Sensor operators sitting together must recognize this limitation and strive to keep ICS communication as clear as possible. If they choose to "side bar," they need to relay relevant information to the rest of the aircrew on ICS after their conversation has come to an end. If they are referencing tactical information, they must use information that is visible and clear to all operators so there is no ambiguity.

Optimizing Information Management

We can categorize ASW aircraft into four groups based on the arrangement of their sensor operates and characteristic of their tactical systems. The first group is aircraft with physically separated sensor operators and fixed function tactical systems. Examples of this type of arrangement include the P-2 Neptune and P-3B Orion. The next group features aircraft with physically separated sensor operators but reconfigurable function tactical systems. This is a rare arrangement, and best describes certain models of U.S. Navy P-3C Orions equipped with Anti-Surface Improvement Program modifications and Royal Australian Air Force AP-3C Orion aircraft. The third group of ASW aircraft feature co-located sensor operators and consoles with fixed tactical functions. A good example is the P-3A Orion. The final group is made up of aircraft that have co-located sensor operators and reconfigurable function tactical systems. The best example is the P-8A Poseidon.

For aircraft with fixed function tactical systems, improving communication is the most effective strategy to optimize information management. Aircrews should use verbiage that is concise, clear, and unambiguous. Standardized phrases and voice reporting formats should be used so that the maximum tactically-relevant information can be exchanged in the shortest amount of time and with minimal misunderstanding.

Aircraft that have physically separated consoles but reconfigurable tactical systems should make use of the communications strategies outlined above. They should also use load-sharing principles as discussed earlier in the chapter to balance tactical tasks in the most efficient manner.

Aircrews who are privileged to operate at physically co-located consoles and with reconfigurable tactical systems enjoy the ideal configuration to maximize the resources of the crew. Aviators can not only assign tactical tasks in the most efficient manner, but also arrange the location of certain crewmembers to maximize information sharing and optimize communication.

Before an aircrew can arrange their operators in the ideal manner, they should first attempt to value stream map the activities conducted during ASW operations. Value stream mapping is a process by which managers conceptualize a production system. They begin by observing the initial state and the desired end state of a system. They then identify the various steps required to achieve the goal.

In value stream mapping the ASW mission, aviators will quickly identify what pieces of data are gathered by each sensor, what steps are required to synthetize this data into tactically relevant information, and how particular pieces of information inform tactical decision-making. Once these various processes are defined, data analysis roles and communications pathways will take shape. Aircrews can then assign tactical tasks to sensor operators in a logical, balanced way and then place them in physical locations that best facilitate communications pathways.

Many U.S. Navy P-8A aircrews fly with a "value stream mapped" sensor operator configuration. During ASW operations, American Poseidon crews fly with five operators at their tactical consoles. The tactical coordinator is responsible for overall employment and supervises the actions of all operators. The communications officer oversees track management, communications with other platforms and ashore installations, and operation of the various communications systems and data links. Two acoustic operators employ active and passive sonobuoys as well as perform TMA. An electronic warfare operator rounds out the crew complement and is responsible for operating non-acoustic sensors, maintaining a plot of surface ships, and providing quality assurance of TMA conducted by other crewmembers.

Operator consoles onboard the P-8A are clustered in a row along the port side of the aircraft in the middle of the fuselage. At the aft end of the "rail" sits the electronic warfare operator. Communication on surface tracks and ELINT flows naturally from the electronic warfare operator to the communications officer sitting on his or her right. The communications officer can easily turn and pass information or tasking that she has just received to the tactical coordinator sitting next to her.

The tactical coordinator inhabits the central position amongst the operator consoles. As he is ultimately responsible for tactical employment, it is natural that he inhabits the focal point of all communication on the rail and be in close proximity to all operators. By sitting in the central console, the tactical coordinator is no more than one seat away from any operator. He can rapidly communicate both verbally and non-verbally with anyone on the rail.

To the right of the tactical coordinator sits the senior acoustic operator. She is responsible for configuration of the acoustic processor and serves as the primary TMA evaluator. The senior acoustic operator is assisted by the junior acoustic operator, who sits at the forward-most console and performs sensor management.

The position of the junior and senior acoustic operators in relation to the tactical coordinator is no mistake. The junior acoustic operator is focused on a "micro level," devoting great attention to the sensor settings, position uncertainty of certain sonobuoys, and configuration of the acoustic system. The senior acoustic operator is focused on a "macro level," sifting through large amounts of sensor data to develop not only an appraisal of the location and motion of the target submarine but also an understanding of the larger tactical scenario.

In the same vein, the tactical coordinator will be developing an even wider mental map of the tactical situation, taking into account not only the position and motion of the target submarine, but also the position of the aircraft, the number and type of sensors remaining, the threat level, and the role of the ASW aircraft in the mission as a whole. In such a configuration, information flows naturally upwards from the junior acoustic operator to the senior acoustic operator and from the senior acoustic operator to the tactical coordinator. Delegated tasks and requests for information flow naturally downwards from the tactical coordinator to the senior acoustic operator and from the senior acoustic operator to the junior acoustic operator.

The physical arrangement of the sensor operators serves not only to facilitate natural communications paths, but also to keep all members of the crew engaged and working at their peak level. Imagine a configuration where the position of the senior and junior acoustic operators was reversed. In such a scenario, if a tactical coordinator needed to ask a question of the senior acoustic, he would have to "talk past" the junior acoustic, bypassing a crewmember to establish a communications link. Human nature and social dynamics dictate that a junior operator in such a position will tend to defer to the more senior members of the aircrew. The junior acoustic will likely draw himself physically back in his seat to allow the tactical coordinator and senior acoustic to make eye contact. He will also tend to withdrawal from the tactical scenario, not wanting to disrupt the important communications between two more senior members.

Aviators use the term "sandbag syndrome" to refer to a crewmember who is dead weight, simply "along for the ride" rather than actively participating in the mission. If an aircrew wants to avoid turning the junior acoustic into a "sandbag," they will avoid placing him in a console location that disrupts natural communication pathways. They will place each operator in a position that makes them feel like a contributing member of the aircrew and encourages engaged, optimal performance.

Voice Communications

In any ASW aircraft, regardless of the console configuration, voice communication remains critical to an aircrew working together effectively. Exchanging information quickly, accurately, and unambiguously is not a simple task. It does not happen naturally. Rather, it requires planning and practice. Communication skills must be developed and sustained both by individual aviators and by aircrews as a whole.

While communicating, operators and pilots have two primary tools in delivering information effectively: their words and their tone of voice. Aircrews should establish standard formats when making tactical voice reports. This way, information is transferred in the same way every time, and helps other crewmembers to expect what to hear next. Hearing various pieces of information in the same order allows crews to more quickly orient themselves to the tactical scenario. They are able to focus on the information itself, rather than on decoding what was said. An unambiguous and standardized report allows other operators to rapidly build a mental map of the tactical scenario and maintain situational awareness.

Tone, inflection, and voice modulation highly influence communication. They can either be either highly effective or highly detrimental depending on how they are used. An assertive voice on ICS can communicate urgency and focus an aircrew on a task or it can communicate anxiety and be distracting. Operators should speak calmly, clearly, and remain positive. They should avoid monotone delivery, as this can be frustrating and grating to listen to. Assertive tones should be reserved for situations where a leader must focus an operator on a critical piece of information or cut through chatter. Excited speech, strained tones, or yelling are almost never productive. They will only degrade situational awareness. If a crew as a whole is becoming excited or agitated, the leader should refocus the crew by using a firm, even, and steady voice on the ICS and by slowing down his or her speech to help calm the crew.

Controlling the Tactical Scenario

Every group or team has a leader. An ASW aircrew is no exception. One crewmember will naturally adopt this role and direct the crew's actions. Generally, this role falls to the most tactically proficient pilot or sensor operator. Tactical proficiency does not automatically correlate with rank or amount of flying time. Varying levels of proficiency result from an aviator's level of ASW experience, interest in the mission, and whether he or she has been flying frequently in the past several months or has been away from the cockpit.

The most proficient or experienced crewmember need not always lead. In the case where a tactically proficient senior pilot is flying with an experienced but more junior tactical coordinator, the pilot will usually defer to the tactical coordinator due to the increased amount of information that can be processed at the sensor consoles as opposed to the tactical displays in the cockpit. Regardless of what crewmember fills the leadership role, it is critical that the aviators recognize who is in charge, let them set the pace, follow their lead, and work together as a team.

Because of the communication barriers posed by physical separation, the relationship between sensor operators and pilots can be challenging at times. Crewmembers must ensure that this relationship remains supportive and collaborative. Communications mistakes will happen, and both sides must be understanding and forgiving of errors. Neither side can allow frustration to lower their situational awareness and degrade their ability to employ their weapons and sensors.

An effective ASW aircrew will attempt to establish an employment rhythm where the pilots and sensor operators are "in-sync." The best aircrews are familiar with the habits of the pilots and the operators. Crewmembers can anticipate the actions of their compatriots. Such a relationship takes time to establish but is very effective and allows a natural employment rhythm to develop.

The best option for an aircrew that does not have this high level of comfort and familiarity is to artificially establish an employment rhythm by verbalizing each task before it is carried out. Whichever operator is the leader will take charge and continuously direct the crew to what task will be accomplished next. For example, a tactical coordinator might direct the pilots to fly north of a target and then come south to employ sensors. Or, the senior pilot might begin to fly the aircraft north of the target while informing the operators that she plans to continue north before turning south and employing sensors. The operators then carry out the necessary tasks. Either of these approaches is effective given humility, collaboration, and a desire of the aircrew as a whole to work together as a team.

The least effective aircrews are those where pilots and sensors operators have an adversarial relationship, or when sensor operators do not follow the direction of the tactical coordinator. An atmosphere like this is toxic. Without a leader taking charge, the efforts of each operator are poorly coordinated, out of sync, and slow. The aircrew is highly reactive and has little ability to control anything but the simplest tactical scenarios.

Just as one member of an aircrew must take charge and drive tactical employment, so too must someone take charge if two aircraft are working together. This is especially important when two aircrews are turning over contact of a target from one aircraft to another. During turnovers, or "swaps," the tactical scenario is taxing. A large amount of information must be exchanged between the on-coming and off-going aircrews. The pilots must operate their aircraft close to the target and close to each other while trying to avoid alerting the target or risking a mid-air collision. The sensor operators must focus not only on conducting TMA and maintaining contact with the target but also on passing information to the communications officer and relief aircrew. These responsibilities lower the situational awareness of an aircrew and make a swap an excellent chance for a submarine to attempt to break contact.

During a turnover, one aircraft must take charge and coordinate the actions of both aircrews. Much like the relationship between the sensor operators and pilots, it does not matter which aircraft leads, just that one of them takes charge and drives the actions of both platforms. Traditionally, Western ASW crews have allowed the on-station aircraft to drive the swap process. This was due mostly to communications and data link limitations that made it difficult for an on-coming aircrew to develop a clear view of the tactical scenario until they were in range of voice communications. As long-range communications and data links have become more common and more capable, on-coming aircraft have become able to develop a view of the battlespace much sooner and much further away from the operating area.

If a relief aircraft is arriving with poor situational awareness in regards to the target, it is usually best to allow the off-going aircraft to drive the turnover. If the on-coming aircrew is armed with good information about the target and the environment, there may be some benefit to them driving the turnover. The on-coming crew is less task saturated than their compatriots and freer to position their aircraft in a supporting manner to quickly employ sensors and take over the prosecution once the off-going aircraft has departed the operating area.

Crew Training and Mentorship

The large number of operators and pilots onboard an ASW aircraft make on-the-job training a natural fit for the mission. Junior pilots and operators have the advantage of watching and learning from their more experienced colleagues in flight. Learning by doing and through osmosis help junior personnel develop skills more quickly. But this learning must be planned and supervised properly so it is effective and does not distract from mission accomplishment.

Teams of pilots are typically built in a manner such that individuals complement and support each other. A new aircraft commander will usually be paired with an experienced copilot. A very experienced aircraft commander might be paired with two inexperienced copilots, since she can carry the load of supervising and training two young officers while still carrying out tactical tasks.

In a similar manner, pilots and operators are assigned so as to complement each other. A new aircraft commander will often be paired with an experienced tactical coordinator, who can run the crew while the head pilot focuses on growing more comfortable leading the junior pilots. By pairing varying levels of experience, squadron commanders balance out their crews, making them safer and better able to complete the basics of the mission.

Senior sensor operators can help introduce their junior compatriots to increasingly challenging assignments to help them develop. For example, a new junior acoustic operator will focus at first on very narrow tasks while the senior acoustic operator controls the tactical system and conducts TMA. As the junior acoustic operator grows increasingly experienced and comfortable, the senior acoustic operator will begin delegating more and more tasks. Eventually, the senior and junior acoustic operators will swap roles during a simple tactical scenario, allowing the junior to conduct primary TMA. The junior acoustic operator will then take on more and more challenging roles, with the senior offering feedback and perspective.

Eventually the senior acoustic operator will allow the junior acoustic operator to fulfill all functions that a senior acoustic operator normally would, stepping in only when a serious mistake is made. In this way, the senior operator has guided the junior through the training continuum and helped them master all the required skills for basic airborne ASW. With more experience, the newly trained junior acoustic operator can take on the senior acoustic operator role in their crew and begin the training and mentorship cycle anew. A similar approach can be used by pilots, tactical coordinators, and electronic warfare operators.

Standardized Tactical Employment

Aviation has long used standardized operating procedures to make complicated and high-risk tasks safer and more efficient. Flight crews follow checklists and air traffic controllers use standard terminology on the radio to communicate the maximum amount of information with the least ambiguity. So too do many military tactical aviation communities use standardized commands, communications, and employment techniques.

An ASW force intent on maximum combat effectiveness should use standardized employment methods as a tool to increase their efficiency and lethality. Used properly, standardization makes individual crews more effective and the force more uniform in its ability to complete the mission. Because every aviator is familiar with basic tactical procedures, personnel can move between crews during training or wartime missions. This flexibility is a force multiplier.

Standardized tactics also allow crews to react more quickly to a tactical situation. Contact with a target submarine may be fleeting. For ASW aircrews, a quick and effective reaction is often the difference between successful localization and a kill or losing the target. In localization operations especially, there is great value in practiced, standardized responses to a particular tactical scenario. These pre-planned responses allow operators to react as quickly as possible and employ effective tactics with a minimum risk of delay or miscommunication.

Using standardized tactics does not require stifling innovation. To the contrary, a standardized force is able to quickly adopt and train to new employment methods. In the same way that standardized flying procedures and communications improve safety, so too do standardized tactics improve aircrew effectiveness.

Aircrews employing standardized tactics must understand the assumptions and entering arguments used to formulate these various pre-planned responses. Sound tactics derive from effective tactical oceanographic analysis and good sensor placement geometry. If crews learn these basic tactics and thoroughly grasp the assumptions in each pre-planned response, they will have a sound understanding of the foundations of airborne ASW. It is only then that aircrews can begin to experiment and form their own approaches to solving various tactical problems.

While sensor capabilities and radiated noise levels change over time, the basic physics of sound propagation in the ocean and electromagnetic radiation in the atmosphere does not. Tacticians have been working to solve search and localization problems for many decades. Before embarking on developing new tactics, aviators should revisit old publications to see if the problem has been addressed previously.

Once a new tactic or procedure is tested and found to be effective, it should be disseminated rapidly. The benefit of standardized tactics is that new procedures can rapidly be adopted fleet-wide, speeding change and improving the effectiveness of not just one aircrew, but dozens of aircrews in short order. It is this characteristic that makes standardization a powerful force to improve warfighting effectiveness.

Experience, Crew Stability, and Tactical Flexibility

There are certain instances where crews may be able to have more flexibility in the manner in which they operate. For example, a crew that is highly experienced and flies together often may like to make minor alterations to standardized voice reports or tactical procedures. This is reasonable, since there are different ways to make a voice report and various ways to fly a sensor employment pattern while producing the same result.

Such tactical variability is not necessarily a bad thing but must be used under the right circumstances. An aircrew that is very stable, with the same operators flying together over a long duration, can afford to use their own style of communication. Another aircrew that frequently has personnel rotating into and out of the team does not have this same luxury. If this constantly changing crew uses communications that are different than those of their squadron mates, a new member will be lost every time he or she flies with them. The effectiveness of the aircrew as a whole will decrease.

The size of a military and its personnel management policies will have an effect on whether less standardized tactical employment is effective or undesirable. In some small militaries, there may be only a dozen patrol aircraft and a few hundred ASW aviators. Because this force is so small and has such a specialized skill set, they are less likely to be transferred to different assignments in other parts of the service. As a result, smaller nations usually have ASW aircrews that enjoy less frequent personnel rotations, have more overall experience, and feature more crew stability. These squadrons can, and often do, adopt more flexible approaches to employment.

Conversely, a large military force, such as the U.S. Navy, boasts many ASW aircraft, hundreds of aircrews, and thousands of ASW aviators. These aviators are often assigned to shipboard duties or assignments that take them away from ASW flying. Because of the high rates of personnel turnover, crews are less stable. Experienced personnel are constantly replaced by junior trainees. This high turnover rate and lower experience level make highly standardized tactics and voice reports very important.

The skills of an ASW force tend to ebb and flow depending on the severity of the submarine threat and the urgency of facing down that threat. At times where ASW forces are in low demand, commanders must give even more attention to ensuring their aircrews utilize solid tactics and standardize their employment. Without an immediate threat, the temptation is great to allow the foundations of tactical oceanographic knowledge and proficiency in basic ASW operations to erode. This strong foundation during lean times will allow an ASW force to more rapidly regain proficiency when these skills are needed. Focusing on fundamentals during times of low demand prevents crews from having to "reinvent the wheel" when ASW again becomes a national priority. In times where the ASW mission is not in vogue, commanders should continuously emphasize the importance of standardized tactics and knowledge of the fundamentals of airborne ASW.

Chapter 12

Search

The airborne ASW kill chain consists of four basic steps. The first of these is the search phase. The process of finding a small object in an area as large as the ocean is a daunting challenge. When one considers that the electromagnetic and acoustic spectrums are cluttered with noise, that the environment changes greatly even over small distances, and that a target submarine may be actively evading searchers, the difficulty of the task becomes clear.

Search is the most difficult part of the ASW mission. It requires planners to analyze a large amount of data, some of which may be time-late and ambiguous. They must work together as a team and optimize the employment of a limited number of search assets. They must coordinate with submarines, ships, and aircraft that are physically separated from each other and operating across vast search areas. This process must not only work in peacetime but also in wartime, when communications pathways are cut or degraded.

To make sense conceptually of the search phase, we will break down the process into five separate steps. First, planners determine what ASW resources are available and what are the capabilities of each platform. Second, planners analyze cueing data and intelligence information to determine where search efforts should take place. Third, planners analyze the detection capabilities of their forces and build a plan to employ each platform. Fourth, sailors and aviators carry out the search operations and monitor the search as it unfolds. As they determine the level of operational effectiveness, they can modify the search plan to improve the probability of detection. Fifth and last, planners abandon the search when detection is no longer feasible and record lessons learned in after-action reports.

Measures of Operational Effectiveness

To properly discuss search planning, we must discuss several concepts used to measure the effectiveness of a search. First is the fundamental relationship between the size of the search area, the detection range of our sensors, and how likely we are to detect a target.

For a given water column and the average AN, sonobuoys will be able to detect a target submarine at some particular distance. If the sonobuoys are spread a certain distance apart, there will be a certain chance of how likely we are to detect the target. We call this measure the "probability of detection." If fewer sensors are used to cover the same area, they must be spread further apart, in turn generating a lower probability of detection. More sensors in the same area will yield a higher probability of detection. In discussing probability of detection, we can break this measurement down further into two terms: instantaneous detection probability (IDP) and cumulative detection probability (CDP).

IDP is the simplest measure of how likely a sensor is to detect a target. It is used for short-duration problems such as localizing a submarine. During longer time periods, such as searching an operating area for several hours, IDP is an inaccurate operational measurement.

To better understand the limits of IDP, let us consider a target submarine patrolling in a square search area. We will assume there is a uniform probability that the target is located at any given point in the box. We will also assume that the ocean environment is uniform and sound propagates with the same velocity at every point in the water column. By analyzing the SL of the acoustic energy the target submarine produces and the AN in the search area, we can determine a detection range for our passive sensors. We can then plot how far from a sensor our target may be detected.

Consider a field of sonobuoys deployed in a grid across the search area. If we assume the sensors are deployed and switched on at the same time, we can model the IDP by determining how likely the target is located within detection range of one of our sensors. We can compare how likely it is that the target is close enough to one of our sensors to register a detection. This measurement of how likely the target is to be detected immediately is referred to as IDP.

However, actual searches are not instantaneous, and therefore any measurement that does not take time into account is limited in reflecting how a search will actually unfold. For example, let us consider the problem of aviators using the same sonobuoy field we discussed above to search an area over the course of several hours. In this scenario, the target submarine is unlikely to stay stationary in the sonobuoy field, but rather, is likely to move around in the search area. The boat may start outside the detection range of a sensor, but it is possible that as the boat moves, it will pass within detection range of one of the buoys.

Modeling search to take into account time, the motion of the target, and the synergistic effects of the sonobuoys working together is known as CDP. The most realistic and effective ASW search modeling utilizes CDP calculations. These calculations can be done for active and passive sensors.

Passive CDP

The most common method of calculating CDP for ASW search uses a modeling technique known as Monte Carlo. In this method, a computer generates hundreds or thousands of notional "tracks," each of which represents a particular pathway a submarine could take while moving through a search area. Modeling software plots the position of each sonobuoy, the detection range around each sensor, and then instantiates the tracks. Over a discrete time scale, the tracks are "driven" throughout the sonobuoy field or barrier and a detection event is registered each time a track passes within detection range of one of the buoys. By averaging these detection events, a probability of detection can be measured for that particular sonobuoy field or barrier in the predicted water column.[1]

In **Figure 18**, we see a notional graph of a CDP curve. Note how the curve begins at a non-zero value at the start of the search. This is due to the fact that when an aircrew deploys sensors, there is a non-zero chance that the submarine will already be within detection range of one of their sensors. From here, the CDP curve moves upwards as time increases before reaching a plateau and flattening out. This flattening effect shows that during the early portions of a search, more time observing the search area leads to large increases in CDP. As the search moves past a certain point, however, increases in time lead only to small gains in CDP.

There are several important nuances to take into account when calculating CDP. First, detection range may not be constant across the entire sonobuoy field or barrier. Some sensors might be in areas of higher AN and some might experience different sound arrival paths due to changes in bathymetry. To be as realistic as possible, CDP calculations should first calculate a realistic detection range at each sensor or at multiple points spread across the sonobuoy field or barrier.

Second, it is important to understand that the position where the submarine tracks are instantiated has a large effect on the CDP calculation. During our example of calculating IDP, we assumed that the target submarine began the scenario inside the sonobuoy field and was equally likely to be located at any one point in the sonobuoy field. From an operational viewpoint, these are not realistic assumptions. CDP calculations should utilize cueing information to define a probability map of where the boat is more or less likely to be. This distribution can be fed into a CDP calculator and used to instantiate the tracks in a more operationally realistic manner.

The actual CDP realized during an ASW mission is likely to vary from the calculated value in preflight planning. Observed AN in the operating area may differ from the predicted values used to calculate CDP, leading to higher or lower search effectiveness. Sonobuoys may drift due to currents in the search area, leading to losses in field integrity. If these effects are observed by tracking AN and the movement of buoys, a CDP calculation can be run post-flight to determine how effective the field or barrier actually was at detecting a boat that might have been nearby. This measurement can then be fed into planning systems to help plan new searches, a technique we will discuss later in the chapter.

Active CDP

Search operations using active sensors can be modeled to generate a CDP as well. Just as is done with passive sensors, notional tracks are generated moving throughout the sonobuoy field. Sensors are considered to radiate at set times, generating a "glimpse" when sound energy may reflect off a target nearby and return to a receiver. Time has a large effect on the ability of active sensors to generate detection events. Increased time observing the search area with active sensors leads to increased CDP just as it does for passive sensors. As a result, search operations with active sensors such as multistatic sonobuoys often follow rigid emission plans, where sound energy is introduced into the water column at set times or in a set sequence in order to closely match the computational modeling performed before flight.

One additional concept to consider is the fact that a submarine operating in a field of active sensors will detect the emissions of the search sensor at a longer range than it can be detected by the sonobuoys. In our discussion of passive CDP, we tacitly assumed that the target would not alter its behavior during the search. This is a much more reasonable assumption in the case when the search sensors do not emit energy into the water column. In the case of active search sensors, however, it is more realistic to consider that the target will be alerted to the presence of the active sonobuoys and the ASW aircraft controlling them.

An evasive target is more complicated to model given that it will probably attempt to use geometry to degrade the effectiveness of the sonobuoys nearby. It is also more likely to use oceanographic features in the water column to frustrate the efforts of searchers. Techniques do exist to model CDP given an evasive target but are more challenging from a computational perspective than simply modeling a non-evasive target.[2]

Radar Search Measures of Effectiveness

Since radar search is a key tactic to be used against conventional submarines, we must define some measures of effectiveness to plan radar search operations. First, we must begin by examining the various conditions that must be met for a radar detection of a submarine to take place. To do this, we can examine the various parameters that combine to determine the Probability of Radar Detection (P_{RD}).

$$P_{RD} = P_R * P_{FOV} * P_{ME} * P_{OA}$$

P_{RD} → Probability of radar detection
P_R → Probability that radar is radiating
P_{FOV} → Probability target is in the radar Field of View
P_{ME} → Probability part of the submarine is exposed
P_{OA} → Probability that operator is alerted

P_R reflects the percentage of the time during the search that the aircrew has their radar activated. P_{FOV} is a measure of how likely a target submarine is to be in the field of view of the sensor. By modeling the radar performance against a snorkel or periscope, aviators can define a detection range for the radar set. The aircraft will be able to sweep a particular distance equal to twice the detection range. By modeling the path that an ASW aircraft flies over the search area, planners can model how likely a submarine is to be within range and field of view at the radar.

In order to detect a periscope or mast, the structure must be exposed above the ocean surface. P_{ME} quantifies how likely it is that the target submarine has a mast raised. For snorkeling operations, this value can be deduced based on the adversary submarine force's doctrine on when they snorkel, how often they snorkel, and how long the snorkeling evolution usually takes. It can also be inferred based on how long it has been since the last time a boat snorkeled.

Last, P_{OA} quantifies the human element of the sensor operator manning the radar station. An alert, rested observer is much more likely to pick out the return from a periscope on a radar screen. A tired, distracted observer is much less likely to make the same detection.

Next, let us examine the ability of an airplane to cover a search area with radar. The term "coverage factor" is used to describe how effective the search aircraft is at observing all parts of a search area. A coverage factor of 0.5 would equate with a search aircraft observing half of an operating area during its time on station. A coverage factor of 1.0 would mean that the aircraft and its radar observed all parts of the search region during the on-station period.

Coverage factor is a good starting point to quantify radar search effectiveness, but much like IPD, is too simplistic to truly reflect the operational situation. A search aircraft will usually patrol an area for several hours. It is unlikely that a SSK will expose a periscope or snorkel for all of that time. As a result, aviators use a metric known as "revisit time" to gauge their performance while searching with radar. Revisit time is the period of time it takes an aircraft to patrol the entirety of the search area and return to a starting point to dwell the radar over that area. Ideally, revisit time should be lower than the amount of time the submarine is expected to expose a snorkel while recharging batteries. This will offer the best chance of detection.

Determining Available Search Resources

Before an operation begins, commanders and planners must take stock of the forces they can use during the search. ASW operations are often very slow affairs, taking place over several days. Planners must recognize they are limited not only by the number of platforms they have, but also the endurance of the crews and number of sonobuoys that are stockpiled.

The location of search areas is an important consideration for land-based ASW aircraft. Higher speed or shorter transits to a search area will make land-based aircraft more effective, as cueing information will be less "time late." The longer the time between a cueing sensor losing contact and a search aircraft arriving nearby, the further the target submarine may have moved, forcing an aircraft to search a correspondingly larger area.

Transit length also influences operational effectiveness once an aircraft arrives at the search area. The longer an aircraft must transit to an operating area, the less time it has on station. Time is a critical factor to maximize the effectiveness of ASW sensors. The more time an aircraft has in an operating area the greater the number detection opportunities and higher the possible coverage factor.[3] Commanders and planners must recognize the unique constraints of the operational scenario and determine how best to balance the time, distance, and force equation given the aircraft, aircrews, weapons, and sensors at hand.

Determine the Location of Search Areas

Even a noisy submarine or one that remains on the surface exposed to detection is a relatively small target to detect in a very large search area.

To give ASW aircraft the best chance of detecting a submarine, the crew must be sent to search in an area with the highest probability of containing the target boat. While discussing the identification of search areas, we will divide our analysis into two basic operational scenarios: that of aviators protecting a friendly vessel from submarine attack and that of aviators attempting to search for a submarine and detect it while the boat is not engaged in attacking a friendly warship.

When an aircraft is conducting escort operations, the problem of determining a search area is simplified. The aircrew is not focused on detecting and destroying hostile submarines, but rather deterring them from attacking and driving them off should they attempt to penetrate a defensive perimeter. The goal is preservation of the friendly vessel or task force and allowing these units to accomplish their mission. If an adversary submarine is detected, attacked, damaged, or sunk while doing so, then so much the better. But aviators must keep in mind that their primary goal is allowing the escorted asset to complete its mission, not to seek out and attack enemy submarines at all costs.[4]

In a broad sense, choosing search areas when escorting a HVU seems to be simple matter. The location of the HVU determines a general area where an aircraft should focus its efforts. However, the choice of where to fly and place sensors is more nuanced. The type of propulsion plant and weapons onboard the adversary submarines will govern what actions an aircrew should take.

The simplest threat to defend against is a conventional submarine armed only with torpedoes. The limited submerged speed and endurance of the SSK means it must approach its target from forward of the beam, inside the LLOA. The relatively short range of a torpedo requires that the SSK launch a weapon from a maximum distance of several nautical miles. These limitations drive an ASW aircrew to focus their efforts in a search area forward of the desired track of the escorted vessel.

A SSN armed with torpedoes is a more challenging adversary due to its high underwater speed and unlimited submerged endurance. Most nuclear submarines have the speed to approach surface ships from all directions. As a result, when defending a HVU from a SSN, aviators must protect all sectors around the HVU rather than just the forward quadrant. A SSN approaching from astern must move faster and generate higher radiated noise levels due to cavitation, flow noise, and machinery noise. The same submarine approaching from forward of the HVU's beam will likely move more slowly and more quietly. An aircraft will have to utilize more sensors ahead of the HVU and fewer sensors behind the HVU to generate equal probabilities of detection in both these sectors. As a corollary, an aircraft can focus more of its effort on a search sector forward of the HVU and relatively less effort astern the vessel to provide a balanced level of protection against attack by a SSN.

A submarine armed with ASCMs is a more challenging adversary for an aircrew. The increased range of these weapons compared with torpedoes means that aviators must detect and confront the threat at a longer range if the submarine is to be prevented from attacking. The location of the HVU and the operational situation allow planners to determine a direction from which an adversary submarine is most likely to attack. This direction is referred to as a threat axis. A threat axis may be influenced by the location of ports, choke points, bathymetry, or topographical features that mask a submarine's approach. Aviators should concentrate their search efforts in areas along the threat axis between the HVU and the expected adversary.[5]

Deterring a submarine is just as critical as detecting a submarine when defending a friendly vessel or task force. A highly visible and overt operating posture is best for this role. Active sonobuoys and radar are the most effective tools to make an adversary submarine commander aware that an ASW aircraft is nearby. Active sensors must be radiated from a position and in such a manner that makes them detectable by a nearby submarine.

When an aircrew is searching for a submarine during independent operations, they must use different techniques to define a search area. Planners must make use of information on the target submarine, adversary operating patterns, and the environment to effectively employ a limited number of search platforms. The most effective manner of doing this is to combine all-source intelligence data to generate a model of how adversary submarines behave.

Submarines tend to operate in different modes during an underway period or a patrol. When a boat gets underway, it generally either departs for a patrol area or remains in a local operating area to conduct training. If the boat is headed on patrol, it usually follows one of several transit lanes towards its patrol box. During transit, the submarine will generally steam at a medium speed, seeking to strike a balance between moving quickly to the patrol area and not moving so fast that it is readily detectable by ASW forces. Once in a patrol area, the boat tends to operate slowly, maximizing the use of its sensors while minimizing the chance it is detected.

By tracking submarines while they are deployed, a navy can build up track information on where an adversary's local operating areas, transit lanes, and patrol areas are located. They can also keep track of position, speed, and depth data to help identify patterns about how a particular class of submarine tends to operate. This information is combined with all-source intelligence to generate what is known as a target motion model. The model describes the manner in which an adversary submarine force deploys its boats and the manner in which individual boats operate. **Figure 19** depicts a notional set of local operating areas, transit lanes, and patrol areas used to develop a target motion model.

Once a target motion model has been developed, planners can use applied mathematics, probability theory, and CAS systems. During the remainder of the chapter, we will divide our discussion of search planning into operations that utilize CAS techniques and operations that are planned manually.

Bayesian Search

The majority of CAS techniques used for ASW planning have centered on the application of a mathematical concept known as Bayesian Inference. Bayesian Inference, or Bayesian search as it is more commonly known, is used to locate an object by first analyzing a series of events that could have led to the object moving to its current location.

Each permutation of motion is assigned a probability weight to describe how likely it was to have occurred. The various permutations with their weights are entered into a computer, which in turn uses the probability weights to simulate various movement scenarios. The computer generates a solution that numerically describes areas where the object is more likely to be located and areas where the object is less likely to be located. This solution is defined by a mathematical function known as a probability distribution function, or "prior distribution" as it is commonly called. The prior distribution is often visually depicted as a probability map, with bright areas depicting areas of high probability and pale areas depicting areas of low probability. The probability map consists of grids with each cell assigned a probability of how likely it is to contain the target.[6]

As a search operation progresses, positive, ambiguous, and negative information is collected and fed into a computer. Just as the initial permutations were assigned probability weights, the confidence levels of information gathered during the search are assigned a weight. Bayes' theorem is used to update the prior distribution with the new information, producing an updated function that describes the likely location of the object. This is known as a "posterior distribution," and is again frequently displayed as a probability map.[7]

As we mentioned before, Bayesian search requires that probabilistic weights be assigned to quantify how likely a boat is to behave in a certain way. This weighting is done using a target motion model. Characteristics like speed or depth can be described as probability functions. For example, historical track analysis may show that a particular class of submarine often transits at a range of speeds. It may also show that that particular type of boat operates at certain depths in the water column. When a search is being planned, cueing information and the target motion model are loaded into a computer. The CAS system assigns probability weights based on the target motion model and generates a prior distribution and probability map that are used for search planning. The graphic output of the probability map is useful to help humans visualize the search problem.

The mathematical output of the system may then be fed into a secondary analysis system, which calculates the most effective way to search given a particular probability function. Using tenets of optimal search theory, computer analysis can generate plans for a single aircraft, multiple aircraft operating together, or several aircraft flying sequential sorties to generate the highest probability of locating the target submarine.[8]

A distinct advantage of CAS systems is that they scale gracefully given the demands of large-scale ASW campaigns. As more and more submarines go to sea, additional permutations can be run, generating probability distribution functions not only of the location of one target boat, but also the likelihood that any one of a dozen boats is active in various areas. Given modern computing power, CAS systems scale in such a way that they can analyze the location of an entire fleet of submarines.

Manual Search Area Definition

Without the assistance of CAS systems, ASW planners must make sense of a large amount of information and do their best to determine a search area. This is generally done by analyzing the geometry of the search problem in order to define AOPs where the target is likely to be located. Once an AOP is defined, searchers attempt to contain its growth and systematically search its confines.

Intelligence on adversary operating patterns plays a key role in manual search planning. Historical track data can be used to identify transit lanes as well as patrol areas. Observations of typical speeds and depths for transit and patrol help to define the growth of an AOP. Efficient means of ingesting all source intelligence and cueing information and a well-run operational plot are critical to maintaining situational awareness about possible locations of adversary submarines.

As a search begins, cueing data is provided to planners. Typically, an AOP is constructed by expanding the geographic extents of the cueing information at a rate equal to the highest likely speed of the target. Typically, this speed is not the maximum speed of the target submarine. Although a nuclear-powered submarine is capable of sustaining its maximum submerged speed for days at a time, it is unlikely to do so because of tactical considerations such as higher radiated noise levels and degraded sensor performance.

Similarly, a conventional submarine is not likely to transit at its maximum speed, as doing so would rapidly deplete its batteries and force it to snorkel, exposing it to radar-equipped search aircraft and increasing its radiated noise level. Therefore, nuclear-powered submarines are likely to transit at a maximum tactical speed limited by radiated noise and sensor effectiveness while conventional submarines are likely to transit at a maximum tactical speed limited by powerplant endurance. These speeds can be observed by analyzing historic track information, if available. Otherwise, intelligence analysts must make assumptions given what they know about the capabilities of a particular class of target submarine.

Suppose planners are provided with cueing information that lists the location and time a target boat was observed. They can define the AOP by plotting a common planning aide known as a farthest-on circle (FOC). A FOC is defined as a circle centered at the last known position of the target whose radius expands at a rate equal to the maximum speed of the boat.[9]

A FOC based on the maximum speed of the target submarine expands very rapidly, often outstripping the ability of ASW forces to contain and search the resulting area. As mentioned above, expanding the FOC at the maximum speed of the submarine is unrealistic given how a target is likely to operate. A more realistic depiction of a FOC is to plot using an expansion rate based on the maximum tactical speed of the boat.

The FOC helps define the boundaries of the resulting AOP. Portions of the AOP can be evaluated as more or less likely to contain the target boat. Geographic and bathymetric features may constrain areas where submarines can and cannot travel. Oceanographic features and bathymetry may create regions that submarine commanders seek out or attempt to avoid. A choke point that restricts the passage from coastal waters to open-ocean patrol areas frequently acts as a funnel. These features act to create zones of high and low probability inside the AOP that searchers should focus on or avoid if they are intent on maximizing their probability of detection. **Figure 20** shows a notional depiction of a FOC, hazard regions, and choke points that serve as natural search areas.

A target submarine need not be located towards the outer edges of a FOC. As we mentioned before, warships often conduct trials and training close to port before departing on operations further afield. They may also leave port to train for several days during preparations for a longer patrol. Planners can use a FOC to describe how far the boat may have traveled given cueing data but should remain aware that the target may be located far closer to the cueing location. A FOC is merely a tool to help define the theoretical boundaries of an AOP, not a sensitive analysis technique to pinpoint the likely location of the target.

Search areas may also be delineated for a force of submarines rather than for an individual boat. During a conflict, adversary submarines must transit through choke points or lanes when leaving port for patrol or when returning from an operating area. Using knowledge of the operating characteristics and endurance of classes of adversary boats and limited cueing data, ASW planners can identify relatively brief time periods or certain areas in the transit lane where boats are more likely to be. They can then concentrate search assets in these high probability areas.[10]

Analyzing Detection Capabilities and Allocating Search Forces

With search areas identified, planners can then move to examine their probability of detection of the target and determine how best to task their search forces. The planners must decide when and where to search with their limited number of platforms, and in the case of patrol aircraft, what configuration of sonobuoy field or sonobuoy barrier should be used.

Earlier in the chapter, we discussed the linkage between search area size and the probability of detection. When search areas are identified, planners must analyze the performance of a sensor at several different points in the proposed search area and then make a rough estimate of CDP for that search region. If the CDP is acceptably high, planners can then do more detailed analysis to develop an optimized sonobuoy employment plan and task an aircrew with a mission.

If the CDP for a proposed search area is too low, planners must make compromises to achieve the desired result. They must either choose to shrink the search area to a more manageable size or raise the number of sensors or aircraft to be employed in that search area. Both actions have the effect of raising CDP. Because acoustic processors typically limit the number of sonobuoys that can be monitored and because aircraft are usually in short supply, planners most often choose to shrink the size of the search area.

Once a single search area or several search areas have been vetted and analyzed, planners and aviators can do detailed planning to determine how best they should deploy their sonobuoys. Algorithms can be used to determine the best location and configuration for the sonobuoy field or barrier. They can also generate optimum "ping plans" for active multistatic systems.

Just as CDP is calculated for a single sortie, so too can it be calculated to model the effectiveness of a group of searchers operating against a target submarine. When combined with CAS systems, planners can generate search plans that optimize the employment of several platforms. This process can also be used for several aircraft flying sequential sorties.[11]

Aviators can also model the detection capabilities of radar and visual sensors. While analyzing the environment, software can be used that provides predicted radar and visual detection ranges. Planners double this detection range to determine a sweep width of each sensor. By analyzing the size of the search area, the flight path the aircraft is likely to fly, the speed of the aircraft while searching, and the sweep width of the sensors, the aviators can determine their coverage factor. This technique is most useful when aviators anticipate the target may snorkel during their time on station or when planning a radar flooding tactic.

Monitoring Search Operations

As search operations take place, planners continue to monitor cueing information, observe the environment, and analyze reports from search assets. Before a search begins, analysis tools will likely have generated an estimated CDP for the planned search given expected environmental conditions. However, variations in the environment and operational factors may cause the actual CDP to vary from the projected value.

As we mentioned previously, variations in AN and sonobuoy drift can alter CDP from the predicted value. Higher than predicted sea states can cause clutter that reduces periscope and snorkel radar detection range, lowering the coverage factor during radar search. These effects can be taken into account to calculate an actual CDP realized during the mission. Planners can use this calculation of actual CDP to measure how likely a target operating in a search area was to evade detection.

For CAS-supported planning, updates of positive contact, ambiguous contact, and negative contact are fed into a computer to generate an updated posterior distribution. Platforms may be tasked to search new areas or employ their sensors in different ways as the search goes on. Just as was done during initial search planning, the posterior distribution can be fed into an optimum allocation of search effort algorithm to determine how best to update the assignments and sensor employment plans for subsequent flights.

Ambiguous information that does not lead directly to localization is useful from a mathematical perspective as well. For example, an aircraft receiving a signal believed to arrive via a convergence zone ray path can assume the target is in one of several concentric annuli. A probability distribution describing the characteristics of convergence zone phenomenon can be fed into a CAS system to develop an updated posterior distribution.

Positive contact that is later lost serves essentially the same function as cueing data. It provides a CAS system with a recent, heavily weighted data point. This can be used to generate an updated probability map as well as recommended employment plans for search assets. Planners should consider the possibility that the submarine they have detected may not be the same target they were searching for. By analyzing all-source information, planners can determine a confidence factor of how likely the detected signal was that of the target. This is especially pertinent during prosecutions where an ASW force is facing off against many adversary submarines.

Negative contact is very useful in providing updates to a CAS system. By using in-situ environmental observations, planners can monitor how effectively an aircrew cleared a particular body of water. Perhaps a sortie was planned to clear a particular search box with 80 percent probability of detection. Higher AN values during the search resulted in a 70 percent CDP for the flight. Planners can input this negative search information into a CAS system. This information helps the search allocation algorithms determine whether an aircraft should revisit that area and search again for the target, or if it is more profitable to look elsewhere.

Taken together, the positive, negative, and ambiguous information generated during a prosecution helps planners effectively monitor and tailor search efforts as the operation unfolds. By constantly updating the prior distribution with new data, a more accurate posterior distribution is generated, improving the chances that the ASW force will find the target.

In-situ environmental information and negative contact can also be used in manual search planning. These measurements can help planners define how likely it was a submarine evaded detection in the area they searched. It is useful for making decisions on whether to continue searching in that area or moving to another location. It can also inform decisions prior to moving a HVU into that search area. Updates to the search plan are much harder to generate, however, given a lack of computational tools.

Transition to Localization

To move further along the kill chain, aviators must transition out of the search phase and into localization operations. When an aircrew detects a signal that may be caused by a target submarine, they should aggressively prosecute the contact. They should always assume that the signal is indeed an adversary submarine until it is proven otherwise. The crew should not disregard contacts they believe to be of low confidence. Bayesian search techniques do not emphasize throwing data away, but rather keeping track of it and assigning it an appropriate probability weight. In a similar manner, crews should be very wary of disregarding an ambiguous signal and declaring that it is not a submarine. Aviators should localize and prosecute a contact until they are sure it is not a submarine.

The desire to aggressively localize does have its limits, however, especially during search operations with active sonobuoys. A crew that is engaged in localization will tend to fixate on several individual sensors rather than monitor all sensors equally, as is typically done during the search phase. For example, let us consider an acoustic operator monitoring a field of sixteen sonobuoys. If the aircrew begins to localize using four sensors of interest, the operator will typically neglect the remaining twelve buoys. Every moment that sensors are neglected lowers the CDP of the search pattern as a whole. Unless aircrews are extremely disciplined about simultaneously monitoring search sensors while concurrently localizing, every second spent localizing will be a second where the aviators are not searching with maximum effectiveness.

In the same manner, the CDP achieved by active sonobuoy patterns relies on injecting a particular amount of sound energy into the water column over a specified time period. This is especially true in the case of multistatic active sensors, which typically use a scripted employment plan, with emissions occurring in a set pattern. If a crew ceases search operations to localize and persists so long that they are unable to complete their search emissions plan, the pattern will not achieve the CDP calculated by planners. Aviators must be cognizant to localize aggressively, but in a manner which does not reduce the CDP of the sonobuoy field below acceptable levels.

Terminating the Search

At a certain point, lack of cueing information or limited search assets may make the prospect of continuing to hunt for the target submarine an excessively expensive affair. After a lengthy time period without cueing information, the AOP may have expanded to such a large area that search operations are no longer feasible with the limited number of ASW platforms. Commanders and planners must realistically appraise the search as it progresses. If the probability of detection is too low, planners must honestly admit this fact to themselves and terminate the search.

Just because a search was not successful in generating contact does not necessarily make it a failure or a waste of resources. Planners should ensure that records of the search are gathered and analyzed. Aircrews and planners should be debriefed candidly and lessons learned transcribed to help improve the search planning and employment process.

Training for ASW Search

Before we close this chapter, it is worthwhile to discuss some considerations about training for ASW search operations. Search is the most difficult phase of ASW flying. To carry all these actions out, an ASW force must be proficient and have deep experience with the mission set. None of this is possible without routine practice. Unfortunately, search is practiced infrequently, and when it is, the scenarios are generally not realistic. The ability to conduct wide-area search is the first skill that dies inside an ASW force.

The difficulty of wide-area search operations and cost of realistic exercises often makes navies unwilling to carry out this type of training in peacetime. Expense is not the only deterrent. Even if an ASW task force employs its assets properly and efficiently, search contains an element of probability that ensures a submarine may evade even the most proficient hunters. Despite this mathematical reality, unsuccessful searches still tend to resemble failures to commanders and seem like wastes of time and resources. This often leads to navies practicing scripted and limited scenarios that bear little resemblance to actual operations. These unrealistic searches are not a suitable way to prepare for combat.

Notes

1. For a discussion of CDP calculations, see Daniel H. Wagner, W. Charles Mylander, and Thomas J. Sanders, *Naval Operations Analysis, Third Edition* (Annapolis, MD: Naval Institute Press, 2000), pp. 131–51.
2. Interview between the author and W. Reynolds Monach, PhD, 11 February 2016. Dr. Monach helped develop various CAS systems for U.S. Navy airborne ASW planning during the Cold War.
3. More time monitoring passive sonobuoys in a particular search are will result in more time for CDP to grow. If non-acoustic sensors such as radar or MAD are being used, more time in a search area will result in higher coverage factor.

4. While the observation that the objective of an aircrew in HVU defense is degrading the ability of a submarine to disrupt the mission of the escorted warship seems elementary, it is worth repeating. This nuance has been recognized by operations analysts but has been neglected by warfighters numerous times in decades past. A focus on offensive submarine hunting delayed the adoption of convoy tactics by the British Royal Navy in World War I. For a discussion of the distinction between offensive spirit and defensive operations, see Jan S. Breemer, *Defeating the U-Boat: Inventing Antisubmarine Warfare* (Newport: Naval War College Press, 2010), pp. 69–78. Also see Arthur J. Marder, *From the Dreadnought to Scapa Flow: vol. 5, 1918-1919: Victory and Aftermath* (London: Oxford University Press, 1970), p. 103.

5. Aircrews should remain aware that an ASCM-equipped submarine must still develop a track of the HVU using organic sensors unless aided by external cueing. This need to develop tracks using own-ship sensors may be the limiting geometric influence in the tactical situation rather than the range of the cruise missiles. William J. Toti, "The Hunt for Full-Spectrum ASW," U.S. Naval Institute *Proceedings*, Vol. 140/6/1,336 (June 2014).

6. For a more detailed discussion of the use of Bayesian search for underwater objects, see Lawrence D. Stone, Colleen M. Keller, Thomas M. Kratzke, and Johan P. Strumpfer, "Search Analysis for the Underwater Wreckage of Air France Flight 447," 14th International Conference on Information Fusion (July 2011), p. 1062. A discussion of Bayesian Inference used during the search for the sunken USS *Scorpion* (SSN-589) may be found in Henry R. Richardson and Lawrence D. Stone, "Operations Analysis During the Underwater Search for Scorpion," *Naval Research Logistics Quarterly*, Vol. 18 (June 1971), pp. 141–6.

7. H. R. Richardson, Lawrence D. Stone, W. Reynolds Monach, and Joseph H. Discenza, "Early Maritime Applications of Particle Filtering," *Proceedings of SPIE*, Vol. 5204 (2003), pp. 165–9.

8. The U.S. Navy began its first foray into CAS applications for ASW aircraft with multiple initiatives in the early 1970s. A representative of these systems was the Computer-Assisted Search and Tracking (CAST) program built for the Naval Air Development Center in 1972. The system used target motion models and recent cueing information to generate a search plan for a single P-3 aircraft. Several years later, various ASW task forces were using CAS systems that could generate search plans for multiple aircraft flying sequential sorties. A good example of such a system is the SEQUENCER and VPCAS analysis tools, which entered service in 1978 and 1983, respectively. By the early 1980s, the U.S. Navy was using CAS systems that could coordinate the efforts of undersea arrays, surface ships, submarines, and aircraft. For details on CAST, see Richardson, Stone, Monach, and Discenza, "Early Maritime Applications of Particle Filtering," p. 172. For more information on the history of U.S. Navy ASW CAS programs during the Cold War, see Daniel H. Wagner, "Naval Tactical Decision Aids" (Monterey: Naval Postgraduate School, September 1989), p. II-58–II-65.

9. Alan Washburn and Ryusuke Hohzaki, "The Diesel Submarine Flaming Datum Problem," *Military Operations Research*, Vol. 6, No. 4 (2001), p. 19.

10. During the Bay of Biscay offensive against German U-boats in World War II, British and American patrol aircraft concentrated on a "ribbon" of water that statistically was most likely to contain German U-boats running on the surface, based on the submerged endurance and speed of advance of boats based in occupied French ports. Despite the lack of computer-assisted techniques, planners were able to use operations analysis to most effectively employ their limited number of long-range patrol aircraft. For details on the ASW campaign in the Bay of Biscay, see Norman Polmar and Edward Whitman,

Hunters and Killers Vol. 2: Anti-Submarine Warfare from 1943 (Annapolis, MD: U.S. Naval Institute Press, 2016), pp. 3–6.

11. Early U.S. Navy tactical decision aids and CAS systems modeled the CDP of a single search pattern deployed by one aircraft. By 1978, the SEQUENCER system began modeling multiple sorties by ASW aircraft to generate a CDP for a longer duration search problems. This capability was featured by subsequent U.S. Navy CAS systems. Wagner, "Naval Tactical Decision Aids," p. II-61.

Chapter 13

Localization

L ocalization is the second step in the airborne ASW kill chain. It is the process by which aviators convert an initial detection of a signal that may be from a submarine into actionable information. Aviators first determine whether the signal is likely to be coming from a submarine. If the signal appears to be a false alarm, they return to the search phase. If the signal is likely to be coming from a submarine, however, they move to establish the position, course, and speed of the boat. If intelligence gathering is the goal, the aviators begin the process of extended tracking. If damaging or destroying the submarine is the goal, the crew works to refine their track of the boat until it is accurate enough that they can attack the target.

In this chapter, we will first discuss how aircrews decide whether a detection event is likely or unlikely to be caused by a submarine and what they do in each case. We will discuss geometric concepts useful during localization and how the localization problem forms an AOP much like that in the search phase. Localization tactics are a sensitive topic, and as a result, we will not discuss them. However, we will continue the chapter by discussing the hallmarks of good localization tactics and how these fundamentals inform localization pattern development. We will discuss the steps aviators take to prepare to localize during preflight mission planning and how they update their tactics in response to environmental changes while in flight. Last, we will discuss four different cases of localization and emphasize unique tactical considerations relevant to each scenario.

Signal Characteristics and Confidence Factors

The first step in localization is to determine if the signal being observed by the aviators is likely or unlikely to be generated by a submarine. We can break down the confidence levels used to evaluate signal characteristics into four categories: certain, high, medium, and low. Let us now consider examples of each category.

In some cases, aviators can be certain that a submarine is nearby. For example, pilots might spot the feather caused by a periscope, fly nearby, and observe the boat just below the surface at periscope depth. Other examples might include witnessing missiles rising from the water when no ships are nearby.

In some instances, aviators might not be able to prove that a signal was produced by a submarine but can nonetheless be confident that a boat is probably nearby. For example, consider an aircrew that detects a signal on several sonobuoys in a search pattern. The tonals are consistent with the target submarine and there are frequency differences between each, showing relative Doppler and therefore an indication that whatever is making these sounds is itself moving. In this case, there is strong evidence that the signal is from a submarine, and the aviators can be confident that they are localizing on a signal of interest.

Another example would be an aircrew witnessing a radar contact suddenly disappear. It is remotely possible that a radar contact could be lost due to a ship sinking at just that moment, but if a patrol aircraft is nearby with its periscope-detection radar active, it is much more likely that the radar contact was an exposed periscope that disappeared as a submarine dove away from periscope depth to evade the aircraft. In this case, aviators can be confident that the contact was a submarine and ought to break off their search tactics and localize aggressively.

In some instances, tactical information might be more ambiguous than the previously two examples and aircrews can only have medium confidence in the signal. Consider a crew that is monitoring a field of sonobuoys in the open ocean with no surface traffic nearby. The crew detects a signal on a single sonobuoy. The signal is a common one, produced by the target submarine but also by several types of surface ships.

In this case, the absence of surface traffic makes it appear as if the signal must be caused by a submarine. However, the aircrew must also consider that this signal is not unique. Furthermore, they know that sounds can propagate for very long ranges under the right environmental conditions. The crew should localize on this contact to determine whether it is caused by a submarine. However, they should remain cognizant that it could just as likely be caused by something other than a submerged boat. This is a good example of a medium-confidence signal.

Let us turn now to the case of a low-confidence contact. Consider an aircrew searching an area with active sensors. After pinging several sensors, the crew receives returns from a location near a shipwreck. The signal does not display any relative Doppler, indicating that the object causing the return is not moving. A passive sensor nearby is within predicted detection range for the target but is not showing any contact. In this case, it is likely that the return on the active sonobuoys is being caused by a shipwreck, not by a submarine. Given this cumulative evidence, aviators would be wise to rate the contact as low confidence and continue searching.

Areas of Probability in Localization

Having discussed confidence levels during localization, we can now turn our attention to examining the geometry of the localization problem. When any sensor detects a target, there is a certain amount of positional uncertainty as to the location of the object that is being observed. Often, the sensors used to search during ASW operations are optimized for long-range detection rather than for positional accuracy. This means that as aviators localize, they must convert this long-range but uncertain position information to short-range, refined position information to establish a track or conduct an attack.

Consider the accuracy and positional uncertainty of the various types of sensors used by aviators during the search phase. For operational cueing, planners use ELINT intercepts that span ellipses many miles wide. They also use sonar arrays that have very long ranges and therefore produce AOPs many miles wide. While aircraft are searching themselves, they use sonobuoys and radar as their primary sensors. Radar is the most accurate of these sensors, and generally gives a target position that is quite accurate. Active sonobuoys plot bearing and range but may have range uncertainty that is quite large. Passive sonobuoys may detect a target at medium range or short range. However, these sensors provide a bearing only, with no range information unless multiple sensors are in contact and a cross-bearing fix is possible.

In each of these cases, the initial detection of a submarine signal can be thought to generate its own AOP. In the case of radar or a visual sighting, this AOP is negligible. However, in the case of acoustic sensors, the AOP can often be relatively large. As a result, there is often a great deal of positional uncertainty early in the process of localization.

By utilizing effective preflight planning, aviators can give themselves an advantage in defining the limits of these AOPs. In the case of acoustic sensors, aviators should already have predicted detection ranges generated preflight. They should also have determined what types of sound arrival paths may be present and if the SSP supports ducting or long-range propagation paths. As the aircraft arrives on station and the environment is sampled, updated predicted ranges can be estimated given the in-situ AN and SSP.

Armed with predicted detection ranges, aviators are well on their way to defining an AOP based off acoustic contact during localization. First, in the case of passive or active sensors, some amount of range or bearing error will be present. In the case of a passive sensor, the AOP can be thought of as a fan extending out from the sensor where the lateral limits are defined by the bearing error and the outer limit is defined by the predicted detection range for the frequency of interest and the in-situ conditions. In the case of an active sensor, the AOP can be thought of as an area defined laterally by the amount of bearing error inherent in the sensor and the range limits defined by the range error inherent in the frequency and waveform of the ping that generated the contact. These two AOP types are shown in **Figure 21**.

Expansion of Localization AOPs

While aviators are localizing, they face one of two possible tactical scenarios. In the first and most advantageous, the sensor that gained contact remains in contact until the aircrew can refine the position, course, and speed of the boat and establish a track. In the second scenario, the sensor in contact loses the signal from the boat before localization is complete.

In the first scenario, the crew works to employ additional sensors to reduce target position uncertainty. This type of localization generally proceeds smoothly and the aviators establish a track without much trouble. In the second scenario, however, the aviators are faced with a much tougher situation. When contact is lost, the AOP defined by the sensor characteristics immediately begins to grow.

Much like the case of a boat during the search phase, the AOP can theoretically grow at a rate equal to the maximum speed of the boat. This is disadvantageous for the aviators, since as we will see later in the chapter, a key tenant of localization is to contain the growth of any AOP and then to systematically sanitize the AOP itself to find the target. An AOP that grows at the maximum speed of a boat can quickly become so large that an excessive number of sensors are required to contain its growth.

As a result, aviators ought to work to contain the theoretical limits of the AOP while localizing. However, if they cannot do this, they should work to focus on the areas of the AOP where the submarine is most likely located. Just as submarines are not likely to transit at their maximum speed during the search phase, so too are they unlikely to evade at their maximum speed during localization. This is especially true if the submarine is not alert to the presence of the ASW aircraft.

Just as we do in search, we can use a maximum tactical speed to define the growth rate of the AOP. This can be based on the technical characteristics of the target submarine or on intelligence information. Should a crew not be able to contain the entire AOP, they should define the high-probability areas of the AOP as an area that grows outwards from the AOP at the time contact was lost at a rate equal to the maximum tactical speed of the boat. They should focus first on containing the growth of the high-probability areas of the AOP. With this complete, the crew can then move to sanitize the interior of the AOP, so as not to offer the submarine a safe haven inside the perimeter where it can loiter undetected.

Matching Localization to AOP Expansion

Let us consider a simple localization scenario to illustrate some key concepts of how an AOP grows when contact is lost during localization. For the sake of this exercise, we will assume that contact has been gained on radar against a periscope but the submarine observes the aircraft and dives. The aircraft is flying at a speed of 300 knots and is 15 nautical miles away. The detection range for passive sonobuoys is 2,000 yards. The maximum tactical speed of the target boat is 10 knots. Each buoy takes three minutes to deploy and descend to depth. Once each buoy is deployed, the acoustic processor takes one minute to ingest information and output useable data to the operators.

Analyzing the situation, we can see that the aviators will lose contact once the periscope submerges. They will have to close the distance to the submarine, fly a localization pattern, wait for the buoys to deploy and descend, and wait for the acoustic processor to yield useful data. The amount of time from when contact is lost until useful acoustic data is available is referred to as the amount of "time-late."

Based on the amount of time-late, the AOP will have expanded to a certain size that the aviators should contain using a localization tactic. The tactic will have to be large enough to contain the entire AOP or at least the high probability portions of the AOP. The larger the pattern, the longer it will take to fly and the more buoys will be used, leading to a larger AOP to contain and fewer sensors for follow-on tracking or attack. Too small a pattern, and the boat is likely to escape.

The aviators analyze the situation and calculate that they will take about five minutes to fly a localization pattern. As they are 15 nautical miles away from the area, they will take three minutes to close the distance. Since we know the buoys take 3 minutes to descend and another minute to integrate on the acoustic processor, we can calculate that the localization tactic will have taken 12 minutes to fly. This means the aircrew can consider themselves 12 minutes "time-late."

In the 12 minutes, the submarine moving at 10 knots will have covered 4,000 yards, or 2 nautical miles. Therefore, the tactic must contain a circular AOP 8,000 yards in diameter. Taking into account the 2,000-yard predicted detection range, we can determine the circumference of the expanding circle and calculate how many sensors are needed to form a ring encircling the AOP. Given that each sonobuoy can detect in a circular area 2,000 yards in radius and 4,000 yards in diameter, we find that slightly more than six sonobuoys would be required to contain the AOP. Therefore, the crew rounds up and decides to employ a seven-sonobuoy circle around the AOP to contain its growth. The field of view of the sonobuoys will overlap slightly more than necessary, but this will offer protection against miscalculation. See **Figure 22** for a depiction of this type of pattern.

The most important concept for aviators to keep in mind is that time-late is based on the time from when contact is lost until the time when the localization pattern becomes useful. Aviators must be skilled at quickly calculating how large and what configuration localization pattern is needed to be contain and sanitize an AOP. The further away from a sensor when contact is lost, the longer the transit time.

It is worthwhile mentioning the time it takes to fly localization patterns. The longer it takes to fly a localization pattern, the longer the time-late value will be. This fact incentivizes aviators to contain the outer edges of the AOP first and then focus on placing sensors in the inner portions of the AOP. In this way, they stop the growth of the AOP but still systematically search the interior.

Fundamentals of Good Localization Tactics

The localization tactics used by aviators during operations are sensitive and cannot be discussed in open source material. However, there are certain fundamental concepts of what differentiates good localization tactics from poor localization tactics. In the following section, we will list seven fundamentals of good localization tactics and discuss each concept in depth. The fundamentals of localization are listed below:

1. Good localization tactics eliminate ambiguity in the type of sound arrival path and thereby refine the rough position of the submarine
2. Good localization tactics contain the growth of the AOP

3. Good localization tactics focus on containing and sanitizing the high-probability portions of the AOP should the entire AOP be too large to contain
4. Good localization tactics are easy to fly
5. Good localization tactics minimize sonobuoy expenditure
6. Good localization tactics terminate in tracking or attacking rather than in extended localization
7. Good localization tactics provide good geometry to support follow-on tracking

The following sections will review each localization fundamental in detail.

Localization Fundamental #1: Good localization tactics eliminate ambiguity in the type of sound arrival path and thereby refine the rough position of the submarine

In some search areas, the combination of target acoustic signature, bathymetry, AN, and SSP will make direct path contact the only type of ray path available for aviators to exploit. However, in more complex environments, aviators may be able to exploit several different types of ray paths and propagation phenomena. If this is the case, aviators must be aware that an initial detection signal could come from one of several arrival paths and should plan their localization tactics to cover each different possibility.

For example, consider an aircrew searching for a SSN transiting in deep water. Intelligence information informs them that their target usually transits deep, well below the mixed layer depth. The SSP provides sufficient depth excess and speed excess to support convergence zone conditions. The boat is assessed to be transiting primarily at medium to high speed and slowing down periodically to observe the environment and come to periscope depth to communicate.

While monitoring a sonobuoy barrier, the crew registers contact on one of their sensors, observing several tonals associated with their target. However, they cannot be sure whether this contact is coming from short-range direct path or long-range convergence zone contact. As a result, the crew elects to employ a localization pattern to take into account direct path contact and then transits out towards the first convergence zone annulus range to employ another pattern to take into account the possibility of the sound arriving via convergence zone. The direct path contact pattern does not reveal a submarine nearby, which helps the aviators by focusing them on localizing on the convergence zone contact. This is a good example of trying to first eliminate ambiguity as to the type of signal arrival path and by doing so, give clues to where the submarine is located.

Localization Fundamental #2: Good localization tactics contain the growth of the AOP.

Earlier in the chapter, we alluded to the importance of containing the growth of the AOP. Aviators may be lucky and remain in contact during the entirety of the localization process. However, if contact is lost, they must move aggressively to contain the growth of the AOP. Failing to do so risks the boat escaping and contact being lost. It also risks using an excessive number of sensors by dropping sonobuoys behind the target and "chasing" the boat rather than employing a barrier on the outer portions of the AOP and then sanitizing the inside of the AOP.

If the initial position uncertainty of the submarine upon detection is low or the time-late is small, the process of containing the AOP is easy. However, this is not always the case. Modern submarines are quiet, leading to short detection ranges and a great number of sensors being needed to contain a growing AOP. Initial detection events may also come in the form of ambiguous or imprecise data, such as a single line of bearing from a passive sonobuoy. In this case, the position uncertainty is large and as a result the localization AOP is large and grows rapidly. It may be tactically unfeasible to contain an AOP of this type. This leads us into the next localization fundamental.

Localization Fundamental #3: Good localization tactics focus on containing and sanitizing the high-probability portions of the AOP should the entire AOP be too large to contain

As mentioned previously, it may not always be feasible to contain the growth of the entirety of an AOP. Much like the search phase, there are usually not enough platforms or sensors to systematically search every part of the ocean. Searchers must identify the areas that have a high probability of containing the target and focus their efforts there. In the same way, aviators should remain cognizant that just because a boat can evade at maximum speed during localization does not always mean that it will. As a result, the high-probability portions of the AOP will expand at a lower rate than the theoretical expansion of the AOP itself.

There are several reasons why a submarine commander might evade at a lower speed than the maximum speed of the boat. First, high speed often leads to very high levels of radiated noise due to loud machinery noise, cavitation from the propeller, and flow noise over the hull. All these sources combine to raise detection range and make the boat more likely to be detected by the localizing aircraft. Second, in the case of conventional submarines, a sprint at high speed can rapidly drain the boat's batteries. Because evasions from patrol aircraft are usually protracted affairs, a submarine commander is unlikely to run down his battery quickly, because this drastically reduces the tactical mobility of the boat for the next day or two.

As a result, aviators should use intelligence reports to help define a maximum tactical speed a boat is likely to use during evasion. Should the aircrew be unable to contain the growth of an AOP during localization, they should identify the high-probability areas of the AOP by expanding the original AOP at the time of lost contact at a rate equal to the maximum tactical speed of the boat. With this high-probability area established, they should move to contain it and then systematically search the interior of the AOP.

Localization Fundamental #4: Good localization tactics are easy to fly

Just because a sensor employment plan for localization is theoretically optimal to contain the AOP does not mean that it is within the performance capabilities of the aircraft. Fixed-wing aircraft must be constantly moving and are limited in how tightly they can turn and how slowly they can fly while localizing. Good localization tactics must fit the speed and maneuverability limits of the aircraft.

Good localization tactics must also be repeatable and able to be flown by even relatively junior, inexperienced pilots. Aviators may have to fly these patterns at night, at low altitude, and in poor weather. They may be fatigued and task saturated with other duties such as monitoring fuel levels or keeping the aircraft away from nearby land. A good localization tactic should be easy to fly and not excessively demanding on the skills of the pilots. It should have built-in tolerance for small errors the pilots may make while dropping their sonobuoys.

Localization Fundamental #5: Good localization tactics minimize sonobuoy expenditure

Localization is not an end unto itself but rather a means unto the end of extended tracking or attacking. As a result, aviators must limit the amount of sonobuoys they expend during localization so they have sensors available to continue to track or to attack. Because modern submarines are so quiet, aviators often must expend dozens of sensors during the search phase just to register an initial detection event and begin localization. By the time the aircrew has completed localizing, they may not have enough sensors left to track for several hours or to hold contact until relieved by another aircraft.

As a result, it is important that aviators strike the right balance between expending too many sensors during localization to ensure contact is maintained and expending too few sensors during localization such that contact is lost. Good localization tactics will use just the right number of sensors to ascertain the position, course, and speed of a target submarine and allow a crew to transition smoothly to tracking or attacking. However, it is important for aviators to not succumb to the temptation of using so few sensors such that the localization process is prolonged. This leads us into the next localization fundamental.

Localization Fundamental #6: Good localization tactics terminate in tracking or attacking rather than in extended localization
In an effort to minimize sonobuoy expenditure during localization, it sometimes happens that aviators lose contact before the localization process is complete. This usually happens because the crews are overly aggressive in driving down the size of the AOP or because the crew are task saturated and "behind" the submarine in their decision-making and sensor placement. In this case, lost contact leads to the growth of a new AOP and another "time-late" situation.

Aviators must not let their desire to employ the perfect localization pattern and save the expenditure of a few sensors be the enemy of employing a good localization pattern that leaves some room for error. For example, consider the case of a crew dropping a circular sonobuoy pattern to contain an AOP. While their calculations might call for a 10-buoy circle, they may choose to move their sensors a bit further out from the center of the AOP and use 12 sonobuoys instead and cover a slightly larger circle.

This tactic does use more sonobuoys but it also lowers the chance that the submarine will slip out of their pattern. Perhaps the boat was moving a bit more quickly than they calculated or perhaps their initial AOP location was slightly off. If the crew tries to use the "perfect" 10-sonobuoy pattern and the boat escapes before they are done localizing, they have now in effect wasted the 10 sensors. Better to employ two more sensors and be firmly in contact and tracking than to use fewer sensors initially, waste the entire pattern of 10 sonobuoys, and then be forced to contain another growing AOP. Aviators should not allow the perfect localization pattern to be the enemy of the good localization pattern.

Localization Fundamental #7: Good localization tactics provide good geometry to support follow-on tracking
As we mentioned previously, localization is the prelude to tracking or attacking. As a result, good localization tactics employ sonobuoy patterns that offer good geometry for acoustic operators to conduct high-quality TMA and accurately track or prepare to drop a weapon. Operators typically

conduct TMA using a combination of bearing information to yield cross fixes and comparative Doppler to deduce relative motion.[1]

For example, let us consider a simple localization scenario where an aircrew is holding contact and localizing on a single line of bearing from a passive sonobuoy. In the first scenario, aviators drop two more passive sonobuoys along the line of bearing, one located down the bearing line at 0.5 times the predicted detection range and the second located down the bearing line at 1.5 times the predicted detection range. Here, the two new sensors gain contact, and we can show their solution to the position of the submarine by plotting their bearing errors and overlaying the area where these two fans of possible contact overlap. The resulting track of where the boat is located is relatively imprecise. This makes it more difficult to conduct accurate TMA.

Now let us imagine the aviators used a different tactic where they drop two sonobuoys, each 30 degrees off the bearing line and spaced at a distance equal to the predicted detection range from the original sonobuoy in contact. Both of the new sensors gain contact and we can plot their bearing fans and areas of overlap where the submarine is likely located. Here, the fans overlap in a much smaller area, and as a result, the TMA for the acoustic operators is much better. **Figure 23** shows a depiction of both localization patterns. This is a simple example of how small changes in the orientation of localization patterns can help to better support tracking and attacking tactics once localization is complete.

Preflight Target Analysis

Before flight, aviators need to study the target submarine so they can best tailor their localization tactics to the situation at hand. Basic search planning will have already determined what is the primary search sensor. It will also have shown the aviators what emissions or radiated noise sources by the target submarine are most easily exploited or are of highest interest.

Aviators should note whether there are any intermittent sound sources made by the target boat that can be exploited at extended ranges. Submarines conduct various noisy evolutions such as running bilge pumps to drain water, running hydraulic pumps to move control surfaces, or running air compressors to replenish tanks. These evolutions may be much louder than the normal sound signature radiated by the boat.[2] If these events happen periodically, they offer unique events that may help the aviators to localize.

The aircrew should study the expected evasion tactics that the target submarine commander is likely to employ. Because of the relatively limited mobility of a submarine and the high speed and high mobility of an aircraft, boats are very vulnerable to prosecution by patrol aircraft. Once aviators "latch on" to a submarine during localization, it is often challenging for the boat to break contact unless there are significant environmental features or other noisy sound sources nearby. As a result, if a submarine crew is aware that an ASW aircraft is localizing on them, they will in all likelihood evade aggressively. Aviators should do their best to understand these tactics so they can take them into account while localizing.

Last, aviators need to analyze the SL of the submarine sound sources of interest to determine whether certain sound arrival paths will be available. Just because the bathymetry and SSP support bottom bounce or convergence zone propagation does not mean the combination of AN and radiated noise levels for particular tonals will allow these arrival paths to be exploited. Analysis of TL curves and Full Field plots will help aviators understand what types of arrival paths they can exploit during search and localization.

Preflight Environmental Analysis

In preparing to localize, aviators should begin by analyzing the search area on a macro level and then focusing on particular zones of the search area on a micro level. They should ask themselves whether the water column is homogenous or heterogeneous in nature. They should note whether certain sound arrival paths are exploitable over the entirety of the search area or if only certain regions are capable of supporting a particular sound propagation phenomenon. They should observe whether there are eddies or fronts that will make localization easier or more difficult. Once the search area has been analyzed as a whole, aviators should move to examine each search and localization sensor in detail.

In the case of acoustic sensors, aviators have many questions to answer while preparing to localize. Analysis of the search area will have shown whether the water column is homogenous or heterogeneous. The presence of eddies and fronts may provide havens for submarines. Aviators should have plans for how they will localize in an eddy or near a front should they register contact nearby.

Aviators should focus carefully on bathymetry and shipwrecks in the search area. Changes in bathymetry will impact what sound arrival paths are available. Pinnacles, seamounts, and shipwrecks can all appear similar to a submarine when observed by active sonars. Aviators should note the location of these objects and take them into account. They should take note of the predicted AN levels in various parts of the search area. If these noise levels change, detection range will change as well, which will impact the geometry of localization patterns.

Environmental analysis is critical for planning radar searches and localization tactics. Aviators should note the predicted sea state and determine how this will impact the detection range of their radar versus a periscope and versus a snorkel. They should analyze the atmosphere and determine whether ducting is likely. If a strong evaporation duct is present, it is possible that radar energy will be trapped and ducted over long distances. This will increase the chance that a boat at periscope depth will detect the ASW aircraft at long range and submerge before the aircrew can approach and detect the exposed masts themselves.

Aviators should analyze the expected tactics of the target boat and its electronic equipment to determine what signals may be of interest for exploitation by ESM equipment. Is the boat likely to radiate its radar or communications equipment? If it does, what is the expected detection range based on signal strength and the atmosphere? Aviators should take note if there are likely to be other radars active nearby that have characteristics similar to the radar employed by the target submarine.

Last, aviators should analyze the environment to prepare for the use of MAD equipment during localization. They should determine what is the normal geomagnetic background noise in the search area. The search zone may be located over a part of the earth with large amounts of magnetic mineral deposits or there may be many shipwrecks nearby. Both of these factors will lead to many false alarms on MAD equipment. The aviators must also examine weather information to determine whether the earth is experiencing temporary geomagnetic activity. Solar storms can cause variations in the earth's magnetic field that makes MAD detection of a submarine very unlikely.

On-Station Environmental Analysis

When aviators arrive on station, they must sample the environment and update their predicted detection ranges and localization plans. First, XBTs should be deployed to measure the SSP. Next, sonobuoys should be used to measure the AN at various places in the search area. This information should be combined to update predicted detection ranges for active and passive sonobuoys at various locations in the search area. The crew should then brief how their localization plans will change at various locations in the search area.

The crew should do its best to observe the sea state and the performance of the radar. Sea state can be challenging to observe directly, but wind information is easily determined using inertial navigation systems and Global Positioning System (GPS) receivers. Wind speed can often be used to infer sea state. By observing outside air temperature during descent, aviators can sometimes detect the presence of ducts that affect radar propagation. However, this is often a time-consuming process that is tough to carry out while the aviators are focusing on other tasks.

The most important action that aviators take while conducting on-station environmental analysis is for them to determine whether in-situ conditions allow them to carry out their planned tactics or if the tactics need to be updated. Often, the environment is similar to what was forecasted and no updates are necessary. However, if significant environmental changes occur, detection ranges will change as well and localization plans may no longer be valid. The aviators must recalculate their detection ranges when they observe significant environmental changes in the search area and should update their localization tactics accordingly.

Four Cases of Localization

At this point, we have discussed the various factors that impact aviators while localizing and covered the actions they should take during preflight planning and when arriving in the search area. Now, we will turn our attention to the four different types of localization scenarios that crews may face. These scenarios each have unique tactical problems that aviators must solve in order to convert the initial detection event into a track of the position, course, and speed of the target. The four cases of localization are found below:

Case I: Sensors remain in contact until a track is established
Case II: Sensors lose contact before a track is established
Case III: Localizing on an intermittent source
Case IV: Localizing on a time-late fix

Case I: Sensors remain in contact until a track is established

Case I is the simplest scenario for an aircrew that is localizing. Here the sensor that detected the initial signal remains in contact while the aviators work to reduce the initial AOP and establish a high-quality track. Because a sensor remains in contact, the crew by definition is not time-late. The size of the AOP at this point is limited to the positional uncertainty generated by the sensor in contact.

For example, consider a single passive sonobuoy that detects a signal of interest. The contact will register as a bearing from the sensor. The lateral limits of the AOP will be defined by the bearing error inherent in the sensor and the extent of the AOP will be defined by the predicted detection range that was updated by the aviators when they arrived in the search area. **Figure 21** depicts this type of passive sensor AOP.

In this scenario, the aviators drop additional sensors near the AOP. This allows them to generate cross fixes, remain in contact, and transition to tracking. Because they are already in contact, they can focus on placing their sensors so that there is good geometric spacing between the buoys and so the sensors support high-quality TMA and allow the aviators to move directly to either tracking or attacking.

Case II: Sensors lose contact before a track is established

We have already alluded to some of the steps required to accomplish Case II localization in our discussion of the growth of AOPs. When an aircrew loses contact during localization, the aviators are faced with an AOP that they must define, contain, and search. First, they must note the initial size and shape of the AOP. This is caused by the positional uncertainty inherent in the sensor that was holding contact. Second, the aviators must determine the speed at which the AOP will expand.

Theoretically, the AOP could expand at a rate equal to the maximum submerged speed of the boat. It is more likely, however, that the submarine will evade at a lower speed. The aviators must choose the likely evasion speed and use this value to define the growth rate of the AOP.

Next, the aviators must take note of how long it will take to conduct four tasks: fly to the location of the AOP, fly and drop a localization pattern of sonobuoys, wait for the sensors to deploy to depth, and wait for the hydrophone arrays to stabilize and the acoustic processor to generate useful information. The time needed to reach the AOP is a function of how quickly the aircraft can accelerate to its top speed and how fast it can fly to the position of the AOP. The time required to fly the localization pattern is variable, based on the configuration of the sonobuoys and the overall size of the pattern. The time needed for the sonobuoys deploy is based on the depth to which they are lowered, and is a set value based on the hydrophone depth selection.[3] Last, the time needed for the hydrophones to stabilize and the acoustic processor to function is based on the construction of the sonobuoy and the capabilities of the processor.[4]

The configuration and size of a sonobuoy pattern is based on matching the detection range of the sensors with the size of the AOP. In order to accurately model detection range for the situation at hand, aviators have to take three main factors into account. These are the likely tactical evasion speed of the target, the likely evasion depth of the target, and the radiated noise levels of the target at the evasion speed. The evasion speed of the boat will generate particular radiated noise levels in several frequencies of interest. These values, combined with the depth of the boat and the depth of the sensor, will result in a particular detection range. Aviators must determine these evasion parameters for the target, take note of the target's acoustic signature at evasion speed, and model accurate detection ranges if they want their localization pattern to perform optimally.

Aviators can also infer certain information during localization that helps them focus on certain areas of an AOP. For example, consider a scenario where an aircrew detects a signal on a passive sonobuoy. The initial bearing is 090 degrees and the crew begins to close with the sonobuoy in question to localize. As they are flying to the area, the signal shows a counterclockwise bearing drift. This means the target is in all likelihood moving in a northerly direction. Before the crew can close, contact is lost at a bearing of 045 degrees.

By examining **Figure 24**, we can begin to make sense of the scenario. The figure shows the initial AOP define by the detection event with the bearing of 090 degrees. Here, the AOP begins to move counterclockwise until contact is lost when the AOP is centered at a bearing of 045 degrees. Examining the figure and trying to interpolate what type of target motion could have caused this detection and then lost contact scenario, several facts become clear. First, in order to generate a counterclockwise bearing drift, the target had to have been moving in a northerly direction relative to the buoy. Second, the contact was probably located towards the outer edges of the detection range. Had the target been moving closer to the sonobuoy, it would not have drifted out of contact to the northeast of the sensor like it did.

Taking these observations into account, smart aviators could gain valuable insight as they planned their next move. Plotting likely paths from initial detection to lost contact shows that the target was most likely moving north. For the sake of simplicity, the aviators could define their AOP as a circular area centered along the 045-degree lost contact bearing and located at one detection range from the sonobuoy. They could expand the AOP based on the assumed evasion speed and focus their localization efforts towards the northerly portion of the AOP, since previous TMA indicated that the target was moving north when contact was lost.

For the sake of simplicity, we have limited our discussion of drawing tactical inference to manual localization planning methods. Readers should note that in past decades, aviators have used computer assistance to define high-probability portions of the AOP. During the Cold War, multiple computer systems were developed for patrol aircraft to guide aviators on where and how they should localize.[5]

Case III: Localizing on an intermittent source
Submarines conduct evolutions where temporary sound sources are generated. These events include firing weapons, running pumps, or bringing machinery up to speed. In certain cases, these events produce noise that is of higher intensity than the normal sound sources that are produced continuously.[6] In Case III, aviators make use of intermittent sound sources to conduct localization.

Aviators can take advantage of the difference in detection range between quieter continuous-sound sources and the higher-intensity intermittent-sound sources to gain valuable tactical insight. By modeling the detection range of a sensor versus each source, aviators can determine rough estimates of how far away a submarine is from a passive sonobuoy for the sensor to register the particular signal.

In **Figure 25**, we observe a visual depiction of the detection ranges of the continuous sources of a boat and a particular intermittent source. If aviators observe continuous sources and intermittent sources together, they can be confident that the submarine is located close to the buoy, since the sensor is detecting the shorter-range continuous sources. Were the sensor detecting only the long-range intermittent sources, the aviators would recognize that the submarine is located at a distance beyond the detection range of the continuous sources.

Perhaps an aircrew monitoring a sonobuoy field detects an intermittent signal caused by hydraulic pumps onboard a target submarine they are searching for. By plotting the bearings and detection ranges of this intermittent source, the crew can determine an AOP based on the ranges and bearing error of the sensors. After two minutes, the source disappears as the pump is shut off. While contact is lost, the crew can simply expand the AOP and begin to localize.

Case IV: Localizing on a time-late fix
In certain cases, the position uncertainty of the initial contact or the time required for an aircraft to close with an AOP may both be so large that containing the AOP is not practical. In this case, aviators can treat the

problem as a small area search problem rather than as a pure localization problem. Here, the crew has a piece of recent cueing information that can help them in employing a small search pattern. Just like the larger search problem, concepts of CDP are very useful here. Aviators may find it effective to plot the initial AOP and expand it based on the likely tactical evasion speed. They can then employ a sonobuoy field and model its performance using CDP calculations. Spreading the sonobuoys apart may not provide an immediate detection, but it is very likely that in the near future the submarine may stumble into the detection range of one of the new search sensors. At this point, the crew can begin localizing using Case I or Case II methodology.

Notes

1. For a discussion of sonobuoy cross fixing and bearing error of passive sonobuoys, see Brian S. Miller, et al., "Accuracy and Precision of DIFAR Localization Systems: Calibration and Comparative Measurements from three SORP Voyages," submission to 65th Scientific Committee of the International Whaling Commission (2014). For details of the directional hydrophone of the SSQ-53 DIFAR, see J.A. Rice, "A Prototype Array-Element Localization Sonobuoy" (San Diego: Naval Oceans Systems Center, December 1990), p. 16.
2. Christopher Norwood, "An Introduction to Ship Radiated Noise," *Acoustics Australia*, Vol. 30, No. 1 (April 2002), p. 21.
3. AN/SSQ-53G sonobuoys are listed as having deployment times of 40, 100, and 180 seconds respectively to descend to shallow, medium, and deep depth settings. "AN/SSQ-53G Directional Passive Sonobuoy," (Dartmouth, Nova Scotia: Ultra Electronics Maritime Systems Inc., 2014), http://www.ultra-ms.com/pdfs/ANSSQ-553G%20Directional%20Passive%20Sonobuoy.pdf (accessed 22 December 2016).
4. SSQ-53G sonobuoys are listed as having hydrophone array stabilization times as long as 240 seconds. Ibid.
5. For example, consider the Search and Localization Tool tactical decision aid built by Lockheed Corp. and Metron Corp. for the stillborn P-3C Update IV. The system was capable of identifying high-probability AOP areas based on the configuration of sensors in contact. It was also capable of providing recommended localization plans to aviators. Daniel H. Wagner, "Naval Tactical Decision Aids" (Monterey, CA: Naval Postgraduate School, September 1989), p. II-45-II-49.
6. Norwood, "An Introduction to Ship Radiated Noise," p. 21.

Chapter 14

Tracking

Tracking is the process by which aviators keep a well-refined plot of the position, course, and speed of a target submarine. The goal is to remain well aware of the motion of the submarine. Tracking can be carried out using one of three basic approaches: close tracking, event tracking, and area tracking. Each of these methods have advantages and drawbacks. Commanders and aviators can choose the method that best meets their goals.

Aviators track a boat for many reasons. They may track to develop excellent TMA before dropping a torpedo in an attack. They might also track to gather intelligence on how the submarine operates or to record its acoustic signature. Aviators might also track simply to stay aware of the operations of a particular submarine of interest.

In this chapter, we will begin by discussing several concepts that are important for good tracking tactics. Just as we avoided discussing detailed and sensitive methods of localization, we will not delve into tracking tactics. Rather, we will review several important ideas so that readers can understand the concepts that aviators use while tracking.

Later in the chapter, we will cover close tracking, event tracking, and area tracking. We will discuss some considerations for each type of tracking method. In the case of event tracking and area tracking, we will review a hypothetical tracking scenario to reinforce some of the concepts pertinent to those two particular methods.

Proactive versus Reactive Tracking

The best aviators strive to remain proactive while tracking a submarine. By anticipating the future actions of a target, practicing good CRM, and dropping sonobuoys at the right time and in the right place, an aircrew can remain in control of the tactical situation. This is much more preferable compared to being caught off guard by a submarine's actions or rushing to catch up because the crew is task saturated and overwhelmed with information.

There are limits to the level of tactical control a crew can achieve, however. As we will see later in the chapter, event tracking and area tracking by their very nature are reactive tactics. In an effort to lower sonobuoy expenditure and stay covert, they cede tactical control to the submarine.

A key tool in remaining proactive during tracking is predicting the future actions of a submarine. There is nothing magical or remarkable about this process. It mostly requires understanding the mission of the target submarine. For example, consider a situation where aviators are tracking a submarine that is assessed to be transiting from its port to a patrol area. The aviators have a general idea about the likely destination of the target based on intelligence reports of past patrols. They know the submarine will have some historic speed of advance that fits with fleet doctrine. They can infer that the submarine will probably spend most of its time at medium to high speed while periodically slowing to observe the environment, employ its sensors, or come shallow to copy a message or transmit to headquarters.

Just as planners do in the search phase, aviators can use target motion models to predict the operating pattern of a submarine. They can analyze historic speed values or take note of the frequency of communications breaks. These predictions are not perfect, but they help aviators conceptualize what actions are likely and what actions are unlikely. This goes a long way to helping an aircrew stay proactive.

Aviators also need to observe the environment and make predictions about how sensors are likely to perform given changes in the water column. Tracking tactics are based on geometric relationships between the target submarine and sonobuoys. Changes in the environment will affect detection range, and this in turn affects the tracking tactics that an aircrew uses. Aviators need to utilize environmental predictions about the position of currents, fronts, and eddies to note when they are approaching a discontinuity in the ocean that will likely impact sound propagation. They need to also monitor AN levels and the SSP and run periodic TL calculations to update their detection ranges. This will allow the aircrew to anticipate and react to environmental changes and prevent losing contact with the target.

Tracking Geometry

To be clear, we will not delve into detailed and sensitive tracking tactics. However, there are some basic geometric relationships in tracking that readers should understand. Two main geometric concepts are down course coverage and cross course coverage.

Before exploring these topics, let us define what we mean when we say down course and cross course. Imagine a submarine moving in the ocean. As the submarine moves, aviators attempt to place sonobuoys nearby to generate cross fixes and conduct TMA. Because the submarine is moving, they naturally place many of their sensors in front of the target so that the boat approaches the buoy, passes by the buoy, and then moves on. This provides the maximum amount of information per sonobuoy.

By down course, we mean the direction that is in front of the line of motion of the submarine. Cross course, on the other hand, is the direction perpendicular to the line of motion of the submarine. While aviators are placing sonobuoys near a target, they try to strike the right balance between placing sensors down course and placing sensors cross course.

Down course sensor coverage ensures that the aviators keep sensors in contact with the submarine. Aviators may attempt to ensure that there are no gaps in coverage between the sonobuoys behind a submarine and the sonobuoys ahead of a submarine. Excellent down course coverage ensures that aviators can keep track of the boat. However, dropping many sonobuoys to do this results in a very high rate of sensor expenditure. Because aircraft only have so many sonobuoys onboard, this can lead to the aviators running out of sensors before their mission is complete.

Cross course sensor converge takes into account the possibility that the boat can turn and provides a buffer in the case of poor TMA. If crews used the minimum of one sensor projected out along the submarine's line of advance, any turn by the submarine would result in the aviators losing contact. As a result, aviators can provide cross course coverage by laying a line of several sensors out ahead of a submarine.

An aircrew can deploy a large number of buoys in a row in front of a submarine to ensure the target passes through this cross course line. However, if the submarine does not turn, these outer buoys are wasted, leading to needless increases in sensor expenditure. Just as with down course coverage, aviators need to strike a balance between ensuring cross course coverage and not wasting sensors.

Plot Stabilization

As aviators conduct TMA, they must account for the fact that all bearing and range information transmitted from the sonobuoys to the aircraft is relative to the sensor rather than to the inertial reference frame of the earth or the aircraft. This presents a challenge, because sonobuoys tend to drift from the location where they initially enter the water. In addition, sonobuoys positioned in a field for search operations or in a line for tracking may drift at different rates due to local differences in currents.

In order to account for the movement of individual sonobuoys, aviators carry out an action known as "plot stabilization." Plot stabilization requires aviators to establish drift rates for individual buoys or groups of buoys. This allows the aviators to conduct TMA relative to an inertial reference frame instead of relative to each sonobuoy.

In the post-World War II and early Cold War periods, aviators typically deployed a smoke canister with each sonobuoy. This allowed the crew to visually keep track of the movement of the sonobuoy by following the smoke. Pilots could overfly each smoke canister and sensor operators could plot the location of each sonobuoy.

As Soviet submarines became quieter and faster, the pace of tracking and rate of sensor expenditure both increased, making the process of marking sonobuoys with smoke canisters unworkable. Aviators began homing in on the radio uplink signals from sonobuoys to mark their location. As the aircraft passed over the sonobuoy, the pilots could plot the location of the sonobuoy or even spot the flotation bag if the seas were calm. This process became known as a "mark on top."

The "mark on top" method of plot stabilization eventually became problematic for several reasons. First, if the ASW aircraft were to mark on top the sonobuoys tracking the target, the aircrew would have to overfly sensors near the submarine, raising the probability of counter detection. Second, variations in currents meant that each sonobuoy drifted at a different rate and in a different direction. The pace of tracking operations against increasingly quiet targets made it unfeasible to mark the location of every sonobuoy. Crews were forced to mark on top one buoy, infer its location and rate of drift, and apply these assumptions to every sonobuoy in the water. This drove uncertainty about the location of sonobuoys, and in turn degraded the accuracy of the submarine track.

By the 1970s, advances in electronics allowed ASW aircraft to use more sophisticated means to estimate the position of the sonobuoys. Equipment onboard the aircraft measured minute differences in the time delay of arrival of the uplink signals from the sonobuoy to the acoustic processor. These sonobuoy reference systems could estimate the position of each sonobuoy in the water without requiring the aircraft to mark each sensor or approach the target location.[1] This allowed the tactical computer to continuously generate an estimate of the position of each sonobuoy and a measure of uncertainty of the information coming from each sensor due to possible position error.

Eventually, the introduction of space-based navigation systems and miniaturized receivers offered the possibility that each buoy could fix its own location in real time. The location could be transmitted in the uplink signal to provide the crew an accurate sensor location and improve the accuracy of the target track.[2] It also provided crews with accurate drift information for each sonobuoy. However, these systems became available at a time when emphasis on ASW was at its nadir. At the time of writing, GPS-equipped sonobuoys are just beginning to be deployed in large numbers.[3]

Revisit Time

Being proactive while tracking requires aviators to think ahead and time their actions precisely. For example, consider an aircrew preparing to drop new sonobuoys ahead of the target submarine. If the crew drops the sensors too early, the boat may turn before it reaches the sonobuoys, meaning that the buoys were wasted. If the crew drops the sonobuoys too late, however, the submarine may already have passed the buoys by the time the sensors descend to depth, stabilize, and send data back to the aircraft.

Time management is a critical part of tracking tactics. It takes time for the pilots to fly the aircraft to the next sonobuoy drop position. It takes time for sensor operators to conduct TMA and for technicians to load new sonobuoys into the launch tubes. It takes time for the sensors to fall to the ocean, descend to depth, and for the hydrophone to stabilize. As a result, aviators are forced to work as a team while they track so that their actions are well timed and that sensors get in the water at the right place and at the right time.

To help the aircrew coordinate, aviators use a term known as revisit time to describe how quickly they can execute the process of flying to a new sensor location, deciding where and when to deploy their sensors, and drop the next set of sonobuoys. Revisit time will change depending on the type of tracking method used. During close tracking, revisit time is very short. During event tracking, revisit time is longer, because crews are conducting slower and more deliberate TMA and employing sensors at longer intervals.

Rate of Sensor Expenditure

To understand why the rate of sensor expenditure is important, it is worthwhile to examine tracking in the context of an entire ASW mission. Patrol aircraft usually work in rotations where several aircraft fly sequentially during a 24-hour or longer prosecution. To successfully accomplish a tracking mission, an aircrew must maintain contact during their event and hand over contact to the aircraft relieving them. If all sensors are expended before the relief arrives, contact will be lost and the ASW force must return to the search phase.

Consider an aircrew that is searching without contact. The aviators may have to expend dozens of sensors in their search pattern alone. When contact is gained, they must drop more sonobuoys in their localization pattern. By the time a position, course, and speed are established and tracking begins, the crew may have used a large proportion of their expendable sensors. They may only have a small number of sonobuoys remaining to track for several hours until a relief aircraft arrives.

As a result, aviators play close attention to sonobuoy expenditure rates. They tailor their tactics to take into account their remaining time on station. Smart crews also account for changes in the environment and the possibility of lost contact. For example, an increase in AN or merchant ships approaching the operations area may cause detection ranges to shrink and sonobuoy expenditure rate to rise. Aviators can look ahead to forecast changing environmental conditions so they are ready for these adjustments. Also, aviators often plan to have a reserve of sonobuoys set aside to localize in case an equipment failure, an error in sonobuoy drop location, or a failed sensor cause contact to temporarily be lost.

Tracking Maneuvering Techniques

We mentioned previously the importance of timing when dropping sonobuoys during the tracking phase. Drop a sensor in the right location too early, and the boat may turn before reaching the buoy. Drop a sensor too late, and the boat may have already passed by the sonobuoy, resulting in a wasted chance to generate TMA. Pilots have an important part to play in the timing of tracking. They use a variety of maneuvering methods while tracking to manage geometry and timing problems.

Maneuvering methods are important because of the aerodynamic limits of large ASW aircraft. Fixed-wing aircraft must fly above a certain speed in order to generate enough lift to stay airborne. Tight turns and banking require additional lift, meaning that pilots must keep the aircraft moving fast enough to be able to maneuver.

Because of the way in which fixed-wing aircraft fly, sonobuoys are usually delivered individually or in lines of multiple sonobuoys. As a result, the necessity to constantly move relative to the submarine and the need to deliver sensors in straight lines lead pilots to fly using several delivery techniques.

One simple delivery method that pilots can use to keep the crew synchronized and managing the timing problem involves turning at the edges of the sonobuoy lines. We will call this technique the 45/225 method, to denote the degrees of turn the aircraft makes during this maneuver. First, the aircrew delivers a line of sonobuoys out in front of the submarine. After completing the delivery, the pilots turn the aircraft through 45 degrees in the direction closest to the submarine. They then reverse their turn, turning through 225 degrees before rolling out on the reciprocal heading of the last sonobuoy drop run.

At this point, the crew is ready to head back inbound to place additional sensors down-course of the target. Consider, for a moment, that the crew needs extra time to conduct TMA or to reload sonobuoy launchers. In this case, the pilots need not worry. They can simply fly straight ahead after dropping the first set of sensors. Once sufficient time has passed, they can initiate their 45/225 maneuver and start back inbound to drop their next set of sensors.

Another useful method involves dropping sensors ahead of the target while flying a racetrack behind the target to position for the next set of sonobuoy drops. First, the aircrew delivers a line of sonobuoys down-course of the target. The pilots then begin a 180-degree turn in the direction of the target. Rolling out behind the target, the crew prepares to drop the next set of sensors down-course of the original sensors. The pilots begin another 180-degree turn, flying out in front of the first set of sonobuoys. They vary their angle of bank to roll out the proper distance in front of first set of sonobuoys to employ new sensors.

The two maneuvering methods we have just discussed are well suited for close tracking, where the revisit time is very small. Perhaps the aviators need to extend their revisit time because they are employing event tracking tactics, which we will discuss in a moment. In this situation, the pilots can utilize a racetrack-shaped holding pattern offset from their sonobuoys to remain near the target but give their crew time to conduct TMA and prepare for the next set of sonobuoy drops.

First, the crew delivers a line of sonobuoys. The pilots fly outbound from the target area and turn 90 degrees to setup a racetrack-shaped holding pattern parallel to the course of the target. Here, they can hold while the crew conducts TMA and waits to deploy their next set of sensors. When the rest of the crew are ready, the pilots can exit the holding pattern and fly inbound to their next set of sensor drop points. **Figure 26** shows these various tracking maneuver techniques.

Counter Detection

All aircraft radiate acoustic energy as they fly. Some of this energy is transferred from the air to the ocean and propagates underwater. As a result, submarines may be able to detect an aircraft flying overhead if the aircraft is noisy and passes nearby. As a result, flying near the target submarine while tracking raises the risk that the aircraft will be counter detected. Additionally, the communications equipment, data links, and sensors onboard patrol aircraft emit electromagnetic radiation that may be detected by a submarine.

Counter detection can pose a serious challenge to an aircrew. Because patrol aircraft are fast and mobile, they pose a great risk to submarines. Once an aircraft has "latched on" to a submarine, it can be very challenging for the submarine crew to break contact. As a result, a submarine commander that detects an aircraft overhead may evade aggressively in an attempt to break contact and regain his or her most important operational advantage: stealth.

A British RAF officer had the following to say to describe the susceptibility of ASW aircraft to counter detection by a submarine:

Another reason for losing the target was counter detection. Like the submarine, the [maritime patrol aircraft] in general, and the U.S. P-3C in particular, made a very detectable and distinct noise if the aircraft was allowed to wander too close or over the top of the target; through its own acoustic systems the submarine would detect us; he could then speed up, slow down or go into a very quiet mode all of which would cause tracking problems. Counter detection became very big later when the targets were much quieter and therefore the buoy pattern had to be dropped closer to get them into contact.[4]

Even if a submarine crew is not able to break contact, counter detection can be undesirable for an aircrew. If the submarine is aware an aircraft is nearby, they may operate in a deceptive manner to try to frustrate intelligence-collection efforts. They may also operate evasively, trying to throw the aviators off their trail. Even if the aviators are successful in tracking, aggressive evasion tends to drive up sonobuoy expenditure.

In some cases, counter detection may be a desirable event. For example, if an ASW force is attempting to deter a submarine, an overt posture may be very beneficial. However, in many cases aviators wish to remain covert while they track and generally do their best to avoid counter detection.

Close Tracking

Now that we have discussed some fundamentals of tracking, we can focus on the three different types of tracking methods. As the name suggests, close tracking involves placing sensors in close proximity to the target submarine. Its benefits include excellent TMA, continuous contact with the target, and excellent tactical control due to the proactive method of sensor placement. The drawbacks of close tracking are the high rate of sensor expenditure due to the large number of sensors employed and the requirement for the aircraft to come close to the target to drop sensors. Approaching close to the target multiple times increases the risk that the submarine will counter detect the aircraft overhead.

The geometry of the sonobuoy patterns employed during close tracking is very important. Aviators work to ensure that their sensors are in constant contact with the submarine. They try to employ sonobuoys to give good bearing cross fixes. They may also try to ensure that there are no gaps in coverage from one set of sonobuoys to the next set of sonobuoys.

Ensuring appropriate down-course coverage can be challenging due to the manner in which submarines radiate noise. Unfortunately for aviators, submarines do not radiate noise in a uniform manner when viewed from above.[5] Because of the physical arrangement of a boat's powerplant and propulsion equipment, sound will radiate in an asymmetric manner. Generally, submarines tend to be quieter when viewed from the bow, slightly noisier when viewed from the beams, and louder when viewed from astern. This will mean that detection ranges generally will be shorter on the bow than they will be on the stern.[6] Aviators must take this into account when conducting TMA and positioning their sensors down course.[7]

As submarines change speed and depth, the detection range of the tracking sensors will change as well. Speeding up will raise the boat's radiated noise levels, allowing detection over a longer distance. If the sonobuoys and target are both below the thermocline, a depth excursion across the layer will lower detection range for sensors below the layer. Aviators must take these tactical factors into account when choosing where to position their sensors.

An American naval aviator described the effect that submarine speed changes have on tracking tactics while prosecuting a Soviet Project 675/ECHO II class SSGN in the North Atlantic in 1970 in the following passage:

> I became scared when after three [sonobuoy tracking pattern] penetrations the ECHO II reduced its speed. The range of the

ECHO II was 12-15 miles, but after the sub pulled back his speed to four knots, the submarine got very quiet! After an hour at this slow speed, the target started back up to eight knots.[8]

Aircrews should also be aware that discontinuities in the water column such as fronts or the boundaries of eddies will radically alter sound propagation. While tracking, aviators should periodically deploy an XBT to observe the SSP. This can help alert the crew to changing conditions. The aviators should also note the position of fronts and eddies during pre-flight environmental analysis and prepare for when a boat crosses into these areas. The disruption in sound propagation can easily cause the crew to lose contact with the target.

Consider an example where a crew might employ close tracking tactics. An international crisis has occurred, and an aircrew is tasked to locate an adversary submarine last detected patrolling near a friendly HVU. The crew detects the target and successfully localizes. Since hostilities have not broken out, the crew does not attack. Instead, they begin close tracking. They develop excellent TMA, good enough to allow them to drop a LWT and have a high probability of kill.

ASW commanders have instructed the crew to attempt to deter the submarine from hostile action against the HVU. As a result, the crew takes steps to be overt in their operating posture. They fly over the location of the target multiple times, making it likely that the submarine crew detects an aircraft overhead. In doing so, they send a powerful message to the submarine commander that any aggressive or hostile action will be met with an attack.

Event Tracking

Event tracking involves spreading sensors further apart and taking note of particular events such as gained contact or lost contact. Rather than cluster sensors tightly together to maintain continuous contact, event tracking spaces lines of sensors at relatively long range. Aviators attempt to predict how the submarine will act in the near future and place sensors at some distance ahead of the boat.

A British RAF sensor operator provided an excellent discussion of event tracking tactics in the following passage:

From there on the crew analysed [sic] the information being received in the aircraft to determine the position, course and speed of the target so they know where to lay the next pattern of sonobuoys. During the tracking of an actual submarine the lead [acoustic operator], the route navigator, the air electronics officer and the tactical navigator would all have to assess the information available to come up with an independent position, course and speed, all within a small percentage of each other to enable a compromise to be calculated for the final solution. Should any one person differ from the others, another conversation would take place to thrash out the reason and hopefully refine the compromise. Should this process falter then the next buoy pattern might go in the wrong place followed swiftly by losing the submarine! The decision was repeated very rapidly, normally every ten to twenty minutes.[9]

The benefits of event tracking include much lower sensor expenditure rates than close tracking. Event tracking allows aviators to fly further away from the target, reducing the change that they will be counter detected. The drawbacks of event tracking include breaks in contact and less tactical control of the scenario due to the more reactive nature of the tactic.

The primary reasons that aircrews choose event tracking tactics are a necessity to avoid counter detection or needing to track using a low rate of sonobuoy expenditure. Event tracking requires very accurate tracking of the target during the periods of contact to determine the course and speed of the boat and to forecast where the boat is headed in the near future. Event tracking also requires deliberate TMA to choose the next sensor position so that contact is regained on the more widely spaced sensors.

To better understand event tracking, let us consider an event tracking scenario. Here, an aircrew has been tasked to search for a transiting SSGN. They have expended many of their sensors laying out a large search pattern. Luckily, the crew detected the SSGN and localized successfully. However, they have several hours to go before their relief aircraft arrives and do not have a large number of sonobuoys remaining. As a result, the aviators elect to employ event tracking tactics to reduce their rate of sensor expenditure and make it to their off-station time with some sensors left over.

The aviators observe the SSGN pass by Post A, a line of three passive sonobuoys laid out across the target's course. As the boat moves near the sonobuoys, the aviators take careful note of the course and speed. Their TMA confirms that the boat is maintaining a medium speed in line with historical norms and seems to be traveling along a historical transit route towards a patrol zone.

The aircrew plots the likely location of the target half an hour from the time that it passes the sonobuoys in Post A. Flying out ahead of the target to this location, they drop a line of five sonobuoys that they call Post B.

Contact is lost a few minutes after the target passes Post A and the crew enters a holding pattern as they wait for the target to approach Post B.

While the crew is holding, they position themselves a dozen miles away from Post A and Post B. This prevents them from overflying the target and minimizes the risk of counter detection. The crew hopes that the "event" they are looking for, namely for contact to be gained on Post B, takes place. However, if contact is not gained at the expected time, the crew must determine what factors changed to cause the lost contact. They must develop a localization plan to take into account each course of action that the target might have taken.

The aviators determine that there are three primary reasons that contact might not be gained on Post B at the expected time. First, the boat could have maintained course and simply slowed down. In this case, the crew elects to wait for more time, cognizant that the target is likely to maintain its overall mean line of advance and will eventually pass through the line of sonobuoys at Post B.

Alternatively, the SSGN could have turned to port or starboard. To cover these scenarios, the crew plans to add two additional barriers of sonobuoys to the northeast and southeast of Post B. These barriers are extended further down course than is Post B, to account for the possibility that the SSGN sped up as well as turned. **Figure 27** shows the configuration of Post A and Post B and the additional barrier sensors to be deployed if contact is not gained on Post B.

Luckily for the aircrew, contact is regained on Post B at the expected time. The crew notes that the target will pass through the northerly portion of Post B. This indicates that the SSGN has turned slightly to port. As a result, the crew begins to plan for their next line of sonobuoys, Post C. In order to account for this turn, the crew positions Post C slightly to the north of Post B. After contact is lost on Post B, the crew flies to the location for Post C, deploys four sonobuoys, and enters their holding pattern to wait. Their lost contact localization plans should contact not be gained at the expected time on Post C remain the same as the plan for Post B. This pattern geometry may be found in **Figure 28**.

Unfortunately, the appointed time of expected contact on Post C comes and goes. The crew recognizes that some tactical factor has changed and they must implement their lost contact plan. They deploy their additional buoys to the northeast and southeast of Post C. They fly towards the area between Post B and Post C, turning on their radar to search for the SSGN.

The crew's lost contact localization plan is successful. As the crew flies in between Posts B and C, the radar operator detects a small contact. A communications mast is spotted on the EO camera, and it appears that the SSGN has come shallow to periscope depth to copy a message from headquarters. The crew elects not to close with the submarine. They keep their distance, attempting to appear like they are searching elsewhere and not reacting to the submarine. However, they note that the submarine still appears to be maintaining its mean line of advance and is heading towards Post C and the additional sonobuoys nearby.[10]

After several minutes, the radar contact disappears and the SSGN heads back to depth. Several minutes later, the crew is rewarded with acoustic contact on one of the sonobuoys in Post C. The aviators continue this process of event tracking until they are relieved by another aircraft.

Area Tracking

Area tracking involves spreading sensors over a widely-spaced area to maintain intermittent contact with a target. By placing sensors far apart from each other, aviators can trade tactical control of the situation for very low sensor expenditure. This allows them to stay aware of a submarine loitering in a particular region without being detected themselves and without having to deploy large numbers of sensors.

As we have mentioned already, the advantages of area tracking include low likelihood of counter detection and very low rates of sensor expenditure. The drawbacks include non-existent tactical control, long periods of lost contact, intermittent contact, and poor TMA. In many ways, area tracking serves more to provide cueing to an ASW force to allow follow-on search operations than it does to maintain a running plot on the position, course, and speed of a target.

However, area tracking should not be dismissed as a useless tactic. For example, should area tracking be carried out using passive sensors, the target boat is likely to be completely unaware that it is being prosecuted at all. This can be highly advantageous to an ASW force. Second, improvements in battery technology and unmanned vehicles may soon result in ASW forces being able to seed adversary submarine patrol areas with long-dwell passive sensors and vehicles and monitor for contact passively from long range.[11]

Consider a scenario where an ASW force may desire to conduct area tracking. A navy is conducting an exercise and a foreign submarine has been active in the waters nearby. It is peacetime and relations with the foreign country are neutral. The ASW force detects the submarine near the edges of an eddy. The submarine appears to be loitering in the eddy, taking advantage of the increased AN and high levels of biological activity to cloak its signature. The boat is relatively old and commanders are familiar with its capabilities.

Rather than prosecute the submarine, commanders elect to task a patrol aircraft to seed the area with two dozen long-dwell passive sonobuoys. Over the next few days, the buoys generate numerous detections of the submarine. While contact is intermittent, the buoy field provides useful data on the location of the foreign boat for very low cost. Commanders are able to keep tabs on a foreign submarine without antagonizing the other country by prosecuting the vessel or by wasting time, energy, and resources in ASW operations.

Notes

1. Sonobuoy reference systems were introduced on U.S. Navy S-3As in 1974 and later in variants of the P-3C. With the removal of ASW equipment aboard the S-3B in 1999 and reduced ASW tasking for the P-3C fleet in the 1990s and 2000s, this equipment fell into disuse and disrepair in the U.S. maritime patrol fleet and its value was not realized by new generations of aviators. Use of these systems would be absent in the U.S. Navy for almost two decades until the introduction of a sonobuoy reference system aboard the P-8A in 2012. For details on the principles of radio interferometry and signal processing, see Sammy D. Stair, "The Applications of the Kalman Filter to the Sonobuoy Reference System on the S3A" (Monterey, CA: Naval Postgraduate School, 1972), pp. 12–19.
2. The British RAF led the way with the introduction of GPS-equipped sonobuoys, with initial operational trials carried out in 1995. These sensors were utilized prior to the retirement of the Nimrod MR2 and cancellation of the Nimrod MRA4. Peter Brown and Trevor Kirby-Smith, "Operational Field Trials of GPS Equipped Sonobuoys," *Proceedings of the 9th International Technical Meeting of the Satellite Division of The Institute of Navigation* (ION GPS 1996) (Kansas City, MO: September 1996).
3. The U.S. Navy was very slow to integrate this technology. Engineering work to integrate GPS into the SSQ-53 sonobuoy began only in 2011 and these buoys were not deployed in large numbers until 2015. "Sparton Awarded $8.9 Million High Altitude Anti-Submarine Warfare (HAASW) Technology Demonstration Subcontract," *DefenseAerospace* (6 September 2011), http://www.defense-aerospace.com/articles-view/release/3/128496/sparton-to-develop-high_altitude-sonobuoys.html (Accessed 24 May 2016).
4. Tony Blackman, *Nimrod: Rise and Fall* (London: Grub Street Publishing, 2013), p. 75.
5. Megan F. McKenna, Donald Ross, Sean M. Wiggins, and John A. Hildebrand, "Underwater Radiated Noise from Modern Commercial Ships," *Journal of the Acoustical Society of America*, Vol. 131, No. 1 (January 2012), p. 92.
6. RAF MR2 Nimrod aircrew discuss the aspect dependence of submarine radiated noise and the effect this has on tracking tactics in Blackman, *Nimrod: Rise and Fall*, p. 75.
7. "Radiated Noise Ranging for Submarines" (Perth, Western Australia: Xenthos, undated), pp. 12–15, http://xenthos.com.au/Ranging%20Paper.pdf (Accessed 21 May 2016).
8. Capt. Edward M. Brittingham, USN (ret.), *Sub Chaser* (Richmond, VA: ASW Press, 2007), p. 150.

9. Blackman, *Nimrod: Rise and Fall*, p. 55.

10. For an account of detecting a modern submarine during an excursion to periscope depth, see the account of a British RAF Nimrod MR2 crew detecting a Project 877/KILO class SSK in the Faroe-Shetlands Gap in 1985. Ibid., pp. 75–6.

11. This approach has been previously pioneered by the U.S. Navy. The introduction of quiet third-generation Soviet submarines led the Americans to develop a long-dwell passive sonobuoy known as the AN/SSQ-102 Tactical Surveillance Sonobuoy (TSS). The TSS was intended to operate for five to seven days at a time, providing intermittent contact of Soviet boats in their patrol areas. The TSS was canceled in 1991 after the collapse of the Soviet Union. Norman Polmar, *The Naval Institute Guide to the Ships and Aircraft of the U.S. Fleet* (Annapolis, MD: U.S. Naval Institute Press, 2005), p. 564.

Chapter 15

Attack

Attacking a target submarine is the final step of the airborne ASW kill chain. In this chapter, we will begin by discussing the factors that affect the probability of kill (P_K) and which of those factors can be influenced by the attacking aircrew. We will then review several fundamentals of attacks. Last, we will introduce and discuss the various steps that aviators must carry out before, during, and after an attack.

Commander's Intent During the Attack

An ASW commander's intent will influence the tactics used while attacking a submarine. Broadly, we can break down airborne ASW attacks into two categories: destruction or deterrence. In the first category, commanders are intent on damaging or destroying the target boat. In order to maximize P_K, aviators will take their time to refine an accurate plot of the target's position, course, and speed. They will analyze the environment carefully to determine how to best deliver their attack and take great care to drop their weapons at the optimum location. This methodical and deliberate approach is necessary to give the best chance of scoring a hit.

In the second category, commanders are more interested in disrupting the target submarine's mission rather than sinking the boat. These types of attacks are usually hasty and reactive in nature. Speed is of the essence, rather than accuracy. In times of crisis, an ASW commander might rather place a warning shot near a submarine to warn it off than attempt to destroy the boat outright. Aviators must understand their commanders' intent so they can choose appropriate tactics.

Outcome of the Attack

Aviators use several terms to describe the effect that their attack can have on a target boat. These three outcomes are a seaworthiness kill, a mobility kill, or a mission kill. A seaworthiness kill occurs when damage from an attack prevents the submarine from remaining afloat on the surface or from being able to remain safely submerged. In the simplest of terms, a seaworthiness kill is what lay observers associate with the sinking of a submarine.

It may be challenging at times for an aircrew to effectively carry out a seaworthiness kill against certain targets. For example, modern LWTs have small warheads that may be ineffective against a robustly-constructed double-hulled submarine. It may take many hits or a very lucky detonation near a vulnerable area such as a propeller shaft seal to cause sufficient damage or flooding to sink the target boat.

Even if an aircrew is unable to damage a target boat enough to sink it, the attack might still cause sufficient damage to prevent the boat from moving under its own power. For example, an attack might cause a target's nuclear reactor to shut down. Without the primary propulsion plant online, the boat would only be able to limp along on the surface on the backup diesel engine, making it essentially immobile and an easy target for follow-on attacks. Alternatively, a detonation near the stern of a boat might cause catastrophic damage to the screw or rudder, preventing the crew from maintaining a desired course. In these cases, the target may still be able to float, but will no longer remain mobile or effective in a military sense. This type of outcome is known as a mobility kill.

The final case, known as a mission kill, occurs when an attack causes a submarine crew to abort its mission and withdraw. For example, shock damage might disable critical equipment such as sonar arrays or weapons control systems. In this case, the boat is militarily ineffective and must return to port for repairs. Alternatively, a credible attack delivered by an aircraft against a hostile boat that is approaching a HVU for torpedo attack might cause the submarine commander to break off and retreat.

Probability of Kill

Aviators intent on damaging or destroying a target submarine must work hard to maximize P_K. When discussing and modeling airborne ASW attacks, operations analysts can break down the factors that influence the likelihood of a successful attack. The figure below lists the five arguments in the P_K equation:

$$P_K = P_{ACQ} * P_{HIT} * P_{FUSE} * P_D * P_R$$

P_{ACQ} → Probability of acquisition
P_{HIT} → Probability of hit

P_{FUSE} → Probability of successful fusing
P_D → Probability of damage
P_R → Weapon reliability factor

It is useful to analyze each of these factors in turn to determine which of them the aviators themselves can influence during an attack. P_{ACQ} describes the probability in a given tactical scenario that the guidance system will detect the presence of a submarine and begin to home on the target. This argument is used in the case of attacks with guided weapons such as torpedoes. Unguided weapons, such as depth bombs or depth charges, by definition lack any type of homing system, and as such, lack a P_{ACQ}.

The significance of P_{HIT} varies depending upon whether the weapon used in the attack is unguided or guided. For an unguided weapon, P_{HIT} represents either the likelihood that a weapon comes close enough to the target for the fusing mechanism to activate or the likelihood that the weapon is close enough to the target to cause damage once it fuses.[1] In the case of a guided weapon, P_{HIT} is the likelihood that the seeker will maintain contact and guide the weapon from the time of initial target acquisition until the time of fusing. The presence and effectiveness of submarine countermeasures, the evasive maneuvers of the boat, and the performance characteristics of the seeker all affect P_{HIT}. In the case of unguided weapons, the accuracy of the aircrew's plot of the target's location has a very large effect on P_{HIT}. If the boat's plotted location is incorrect, depth charges deployed by the aviators will stand little chance of detonating near the target.

Once a weapon has come very close to the target, has impacted the target, or has reached its fusing point, the fuse must function properly and detonate the warhead. P_{FUSE} is the likelihood that the fusing mechanism will operate correctly. It is a function both of the reliability of the sensing mechanism that initiates fusing action and the mechanical and electrical reliability of the fuse.

The next factor, P_D, represents how likely the weapon is to cause damage to the submarine. This factor can vary greatly, depending on the type of weapon, the construction of the target submarine, and where the weapon detonates relative to the boat. It is influenced by the size and type of warhead and the manner in which a guided weapon approaches the target before detonating. The hull construction, internal layout of the target, and damage control capabilities of the submarine crew also affect the boat's vulnerability to attack.

The final factor, P_R, models the mechanical reliability of the weapon itself. Due to design factors, manufacturing defects, internal degradation, and maintenance errors, some percentage of weapons will fail to operate properly when deployed. P_R is a function of the design of the weapon and the maintenance and storage procedures used by support personnel. It cannot be affected by the aviators themselves.

Of the factors listed above, P_{ACQ} and P_{HIT} are the only factors that can be directly influenced by the aircrew during flight. The other factors are all the result of the design, construction, and maintenance of the weapons themselves and the design of the target submarine. Aviators may influence weapon characteristics in peacetime by helping to define weapons requirements and driving the acquisitions process, but once aloft, their influence is limited to how accurately they can track the target, analyze the tactical environment, and deliver their weapons.

Before delving into further detail, we must be clear about the differences between delivering guided and unguided weapons. In the case of delivering a guided weapon, aviators must place the weapon in a position where its seeker head is likely to acquire and guide the weapon to impact. This means the aviators are focused on maximizing P_{ACQ}.[2] In the case of an unguided weapon, the aviators must do what they can to get the weapons to enter the water, sink to depth, and come close enough to the target to fuse, detonate, and damage the boat.[3] This means the aviators must analyze the movements of the target, predict where to deliver their weapons, and deliver accurately to maximize P_{HIT}.

Maximizing the Probability of Successful Attack

The first step in maximizing the probability of successful attack is for planners and aviators to select the best weapon for the environment and target at hand. For the last half-century, the two types of weapons used to airborne ASW attack have been guided LWTs and unguided depth bombs. For mobile targets, the ability to guide generally makes the torpedo the optimum weapon. However, for a target such as a bottomed submarine or a slow SSM operating in very shallow water, depth bombs might offer a much higher probability of successful attack.

Regardless of whether guided or unguided weapons are used, aviators must primarily concern themselves with the accuracy of their TMA. This is the single greatest factor that will influence the P_{ACQ} in the case of guided weapons and the P_{HIT} in the case of unguided weapons. If the aviators are unable to maintain a highly accurate plot of the position, course, and speed of the target submarine, they will in all likelihood not be able to deliver their depth bombs or torpedoes to the appropriate location to hit the target or allow the torpedo seeker to acquire.

Luckily, refining TMA for attack is an outgrowth of tactics used during the tracking phase. As aviators prepare to attack, they refine their plot to achieve what is known as a "weapons quality track." A weapons quality track is one accurate enough that the aviators can be confident their weapons will fuse close to the boat or the torpedo seeker is likely to detect the boat and begin to home.

Plot stabilization is a very important part of achieving a weapons quality track. As we mentioned earlier, all information that is transmitted to the aircraft from the sonobuoys is relative to the sonobuoys as opposed to relative to the inertial reference frame of the earth or the aircraft. As a result, in order to refine TMA, aviators must have high certainty as to the position and drift of their sonobuoys. If sonobuoy position information is uncertain, so too is the track of the target.

Once the aviators have refined a weapons quality track, they must determine what is the best location or locations to deliver their weapons relative to the target. In the case of depth bombs, aviators generally attempt to deliver their weapons in a pattern that covers the area where the boat is likely to be located. If the weapons have variable depth-fusing settings, the aviators will have to take the underwater motion of the depth bombs into account. They will have to adjust how far ahead of the boat they must aim, to account for the delay in the weapons sinking to depth. They must also select the appropriate fusing setting so that the weapons fuse at the same depth as the submarine.

In the case of a LWT, the aviators must determine where relative to the target they should deliver the weapon. When a torpedo enters the water, it generally swims to an activation depth and then begins a preset search pattern, utilizing passive or active sonar to detect the presence of a target. The weapon takes time to descend to activation depth and run the initial phases of a search pattern, during which time the target continues to move.

The detection performance of a torpedo seeker is not uniform. Environmental factors such as SSP, the presence of thermocline, AN level, and the presence of shipwrecks or bottom features all affect seeker performance. So too does the aspect of the target with respect to the torpedo and the range from the target to the torpedo. As a result, determining the ideal location for aviators to deliver a LWT relative to a target is a complicated process that is dependent on a host of factors.

Because of the variety of factors at work, engineers and operations analysts generally conduct detailed modeling of torpedo performance in laboratories to develop weapons employment guidance. Aviators are provided with guidelines that help them determine the azimuth and range relative to a target where they should deliver the weapon, the depth at which they should command the torpedo to activate, and the type of search pattern the weapon should be commanded to run.

Torpedo seeker performance modeling is computationally intensive. Engineers must analyze a variety of environmental factors and operational scenarios for each weapon and target submarine class. Due to the computational demands of these tasks, this modeling has typically been accomplished in laboratories. However, the variability of the environment makes it unlikely that the modeling used to develop weapons employment guidelines will precisely match the conditions that an aircrew observes while carrying out an attack. This makes preflight weapons employment modeling inherently limited.

The increased processing power available aboard ships, submarines, and aircraft may change the manner in which weapons modeling is conducted. In the future, aircrews might have modeling software embedded in their tactical systems that takes into account in-situ environmental conditions, bathymetry, and target submarine information to generate weapons employment guidelines that are tailored to the exact tactical scenario at hand. This capability would greatly enhance an aircrew's ability to accurately gauge P_{ACQ}.

TMA Quality

The quality of TMA that an aircrew needs prior to releasing a weapon depends on the intent of the attack. If the aircrew is attempting to damage the target, they must make every attempt to score a hit on their first shot. A submarine may not be alerted to the presence of an ASW aircraft before the attack, but once the torpedo enters the water, the target submarine will evade aggressively and the quality of TMA will in all likelihood quickly degrade. Aircrews intent on scoring a seaworthiness kill or mobility kill ought to refine TMA to the best of their ability before they begin their attack.

In the case that the attack is meant only to deter or drive off the submarine, P_{ACQ} is not as important. In some cases, such as a deliberate warning shot, a weapons hit may not be desired at all. As a result, TMA accuracy is not as important and a weapons quality track may not be needed. The most important action for an aircrew in this type of scenario is to deliver a credible attack in a short amount of time.

Submarine Evasion

A HWT delivered by a submarine or a LWT delivered by a warship in an over-the-side attack must run a relatively long distance from the shooter to the target before impact. This means that a target submarine will often be alerted by the sound of the approaching torpedo and take evasive action to prevent the seeker from acquiring and homing.

Air-launched torpedoes, on the other hand, may be delivered in very close proximity to a target submarine. This makes an airborne LWT attack very threatening to submariners. Because of the relative vulnerability of submarines to attack, crews must take a credible threat of torpedo attack very seriously. ASW aircrews must be cognizant of the fact that as soon as they deliver their weapon, the target will evade aggressively.

During an engagement, a submarine crew is likely to use a combination of passive countermeasures, active countermeasures, and maneuvering to degrade both the performance of the inbound weapon and the TMA of the aircrew overhead. Active countermeasures such as mobile decoys may present alternate targets that closely mimic the signature of the submarine itself. Passive countermeasures may create enough broadband noise that passive TMA is impossible for acoustic operators.

These evasive actions are aimed primarily at lowering the P_{ACQ} and P_{HIT} of the LWT by reducing the chance of successful initial acquisition and subsequent homing. However, they also have the secondary effect of making it difficult for the aircrew to maintain sufficient TMA to deliver a subsequent attack should the first weapon miss. In some cases, the cacophony of noise once a torpedo enters the water may even lead to lost contact. Taken together, these factors make it imperative that attacking aircrews make every attempt to make their first weapon count and continue to track during the engagement.

Steps of the Attack

We can break down the attack conceptually into eight steps, which are listed below:

1. Ensure attack is authorized
2. Ensure weapon has a suitable P_K
3. Preset the weapon
4. Determine appropriate weapons delivery point
5. Determine a plan for follow-up attacks

6. Deliver the weapon

7. Monitor weapon performance and conduct bomb hit assessment

8. Deliver a follow-on attack or execute lost contact tactics

In the following sections, we will cover each step in turn.

Step 1: Ensure attack is authorized
The history of ASW has is replete with instances when ASW forces have delivered attacks against friendly submarines. In order to prevent fratricide, an aircrew must be certain their weapons are de-conflicted with friendly submarines and that they have permission from commanders to carry out an attack.

In order to de-conflict friendly submarines and friendly ASW forces, navies utilize schemes to manage waterspace. For example, an ASW commander might allocate an operating box to a SSN and direct that no other friendly forces be allowed to carry out ASW attacks in that region so as to prevent the friendly boat from coming under attack. Alternatively, commanders might restrict friendly SSNs from operating outside their areas and declare that any target outside the friendly submarines' operating area may be attacked without delay.

Identification and classification of a submarine target also pose challenges to ASW aviators. Acoustic data may be ambiguous. An aircrew might be able to state with confidence that a submarine is nearby but might not be able to precisely identify the submarine based on the acoustic signature. Given the fact that various nations operate the same class of submarine, it becomes clear that appropriate procedures to classify and identify targets must be in place before a weapon is delivered. Navies use intelligence information and acoustic databases to help carry out these tasks.

Step 2: Ensure the weapon has a suitable P_K
Aviators must ensure the weapon they are attacking with is paired appropriately with the target. Ideally, planners have already chosen and assigned an appropriate weapon during preflight based on the likely target class and environmental conditions. In any case, an aircrew should analyze the in-situ environmental conditions, the target submarine, and the weapon's capabilities to determine whether the attack scenario has a suitably high P_K to make expending a weapon worthwhile.

ASW weapons in general are highly specialized, expensive, and few in number. Historical analysis has demonstrated that the expenditure of ASW weapons in combat tends to be higher than expected due to operational factors such as false contact, equipment failure, and poor operator training. As a result, it is critical that aircrews not employ a weapon when the P_K is too low to realistically meet the commander's intent.

An exception to this guideline is in the case of attacks for deterrence purposes. In this case, an aircrew desires to drive the submarine off, disrupt its attack, or send a message to the adversary commander. Here, speed and the credibility of the attack are crucial. Because a kill is not required, or in some cases, not desired, a low P_K is not a reason to break off the attack.

Step 3: Preset the weapon
ASW weapons have a variety of settings that may be manipulated by aviators to tailor the performance of the weapon to the tactical situation at hand. Unguided weapons such as depth bombs have depth and fuse settings. LWTs allow a variety of seeker run patterns and fusing options to be selected. Aviators must take into account TMA information, environmental information such as SSP, and seeker performance versus a particular target class while applying presets to a weapon. Inappropriately assigned presets can radically degrade the effectiveness of a weapon. As a result, it is very important that an aircrew gets this critical part of the attack right to ensure that P_K is as high as possible.

Step 4: Determine the appropriate weapons delivery point
The point where the ASW weapon enters the water relative to the target has a very large effect on the P_K of the attack. In the case of unguided weapons, an aircrew must first develop an accurate plot of the target's position, course, speed, and depth. They then must deliver the weapons to the appropriate point to compensate for the motion of both the target and the weapons underwater. Last, they must ensure that the weapons are set to detonate at the appropriate depth. In the case of guided weapons, the aircrew must deliver the LWT to the appropriate location relative to the target to allow the weapon to swim to depth, activate its seeker, and run its search pattern.

As mentioned previously, aircrews rely on weapons delivery information developed by engineers during development and testing. This information is often referred to as "splash points," denoting the position relative to the target submarines that weapons should enter the water to achieve the desired effect.

Since World War II, ASW aircrews have delivered torpedoes and depth bombs in ballistic flight paths from very low altitude. This was done to minimize errors in plot stabilization and the effect of unguided ballistic flight before water entry. As a result, aviators define a splash point and then use ballistic fall tables to determine the point in space where the weapon should be released from the aircraft.

During the last decade, the increased threat of submarine-launched SAMs and the desire to monitor sonobuoy fields over larger areas have led to the development of wing and guidance kits for LWTs. These systems allow LWTs to be dropped from low and medium altitudes and glide for several miles before entering the water. This tactical scenario requires aircrews to take into account the kinematics of a weapon in flight before water entry. Deploying from altitudes greater than 10,000 feet and at ranges of several miles means that a weapon will be in the air for several dozen seconds at least or perhaps longer than a minute. During this time, the target continues to move. As a result, aviators must take TMA into account and ensure that even after an extended flight, the LWT arrives at a location relative to the target where the seeker is likely to acquire and home to impact.

Step 5: Determine a plan for follow-on tactics

Ideally, an ASW aircrew delivers an attack and the first weapon impacts the target and achieves the desired kill. However, aircrews must consider the possibility that their attack will not be successful and prepare to regain the tactical initiative and continue attacking. If an attack fails, an aircrew will be faced with two scenarios. In the first scenario, the aircrew is able to continue to conduct TMA during the attack. In this case, once it is clear that the first attack has failed, the crew should move to refine TMA to a weapons quality track and attack again. In the second scenario, the crew has lost the ability to maintain track of the target due to factors such as countermeasures, evasive maneuvers by the target, or high AN caused by explosions or countermeasures in the water. In this case, the crew must deploy a lost contact pattern and begin to localize using the last known location of the target as a datum.

Step 6: Deliver the weapon

Analyzing P_K, determining appropriate weapons settings, and plotting a desired weapons delivery point are generally the responsibilities of sensor operators. Once these actions are completed, delivering the weapon accurately is primarily the responsibility of the pilots. The pilots must fly the aircraft to the appropriate point in space at a time designated by the sensor operators and ensure that the aircraft is maintaining the correct altitude, airspeed, heading, and load factor at the time of weapons release so that the weapon travels the appropriate ballistic path before water entry.

The coordination of timing and aircraft placement is especially critical during weapons delivery. Often the TMA necessary for a weapons quality track is based on events such as the target submarine passing by a sonobuoy or the detection of the target with MAD equipment. As time passes from these events, the track quality and reliability of the TMA decreases. As a result, the pilots must coordinate closely with the sensor operators so the aircraft is in the appropriate position relative to the target and reaches the weapon release point at the correct time.

Good ASW pilots do their best to keep their maneuvers smooth and gentle in order to reduce the chance that sensor operators get disoriented or airsick. The pace of airborne ASW is such that it is usually best for the pilots to fly conservatively rather than throw the aircraft around the sky and make the rest of the crew miserable. However, maneuvering to execute a timely and accurate attack or prevent lost contact are exceptions to this rule. When the situation calls for it, pilots must maneuver as needed to put the aircraft in the appropriate position and get the weapon into the water at the right place and at the right time.

Step 7: Monitor weapon performance and conduct bomb hit assessment

Once a weapon has been delivered, the aircrew must monitor the weapon, the target, and the environment as the engagement unfolds. In the case of attacks with LWTs, the kinematics of the scenario may be such that the weapon takes several minutes to deploy, search, and home. All the while, the submarine will likely be evading aggressively. The aircrew should do their best to determine whether the weapon is functioning normally, conduct TMA and continue to track, and discriminate between false targets caused by countermeasures and the signature of the target itself.

When a weapon fuses, the sensor operators will in all likelihood detect the acoustic signature or visual signature of an explosion. However, analyzing whether the weapon caused damage to the target is a much more complicated manner. The information available to an aircrew after an attack is usually ambiguous. In the confused tactical environment, TMA will likely be poor, the acoustic frequency spectrum will probably be cluttered with noise from countermeasures, and mobile decoys may be present.

Aviators have very few telltale signs that the weapon homed to impact and damaged the boat. For example, consider a situation where an aircrew attacks a target, the weapon explodes, and the sensor operators hold no passive contact but observe a stationary target underwater on active sonar. Has the submarine been disabled or is it merely sitting still and "playing dead?" Alternatively, consider an aircrew that attacks, observes an explosion, loses all active and passive contact, but observes debris and an oil slick reach the surface 20 minutes later. Has the boat been sunk? Or might the crew of the submarine have set an ultra-quiet mode, begun to drift, and released equipment and fuel in an effort to fool the aviators overhead?

Such tactical scenarios seem fanciful, but both have historical precedent. German U-boats in World War II used similar tactics on several occasions. It is important to remember that aviators can determine whether weapons have fused but have few concrete clues at their disposal to precisely determine what sort of damage has been done to the boat. As a result, aircrews are best served by using as many pieces of tactical information as possible to determine the status of the target. They should continue to prosecute until they have a high level of certainty that contact is lost or the boat has truly been sunk or disabled.

Step 8: Deliver a follow-on attack or execute lost contact tactics

In the aftermath of an unsuccessful attack, an aircrew will either have maintained contact during the engagement or have lost contact. In the case that aviators retain contact, they should continue to conduct TMA as the attack unfolds. If the aircrew still retains a weapons quality track, they should move to deliver another weapon. If the aviators maintain contact but do not have a suitably accurate track, they should refine TMA until a weapons quality track is obtained and then deliver a follow-on attack.

If the aviators lose contact during the attack, they should move to execute lost contact tactics. They should define a datum using the last known location of the submarine, define the growth of the AOP based on the target's expected evasion tactics, and move to contain the AOP's growth, if possible. Like any other localization scenario, they should attempt to regain contact and transition to direct path tracking. They should then conduct TMA to achieve a weapons quality track and deliver a follow-on attack.

Notes

1. This stipulation is unwieldy but is written in such a manner as to represent differences in the fusing of depth charges and depth bombs. Conventional depth charges are equipped with a hydrostatic fuse so as to detonate at a set depth. Less frequently, they may be equipped with an influence fuse that detects the presence of the boat nearby and detonates. Nuclear depth charges are simply set to detonate at a set depth via a hydrostatic fuse.
2. Unlike submarine-launched HWTs, which generally are wire guided, air-dropped LWTs are independent of the aircraft once launched. As a result, once the seeker acquires a target and enters homing, aviators cannot influence the weapon any further. P_{HIT} is driven by the performance of the seeker and its ability to maintain contact, reject the presence of countermeasures, and provide guidance commands to the weapon.

3. For many years during the Cold War, NATO maritime patrol aircrews carried nuclear depth charges. The very large explosive power of these weapons meant that aviators needed only a rough plot of the target's location. As long as a weapon was delivered in proximity to the submarine, a seaworthiness kill was very likely to occur. For details of nuclear depth charges, see Norman Polmar and Edward Whitman, *Hunters and Killers Volume 2: Anti-Submarine Warfare from 1943* (Annapolis, MD: Naval Institute Press, 2016), pp. 171–2.

Index

Absorption, 98, 116, 126–7, 131, 142
Abyssal plain, 153
Acoustic decoy, 89–90
Acoustic energy, 14, 61, 88–9, 94–5, 97, 100, 104, 115–17, 120, 123–4, 126–7, 129–30, 132, 142, 147, 188, 230
Acoustic impedance, 119, 127
Acoustic Intelligence (ACINT), 155, 163
Acoustic jammers, 11
Acoustic operator, 176, 179, 183–4, 201, 233
Acoustic processor, 95, 97, 147, 179, 209, 219, 226
Acoustic quieting, 73
Active CDP, 190
Active sonar, 4, 28–30, 33, 38, 50, 53, 68, 73, 75, 84, 88–90, 94, 100–03, 105, 111, 114, 136–7, 147, 242, 248
Afternoon effect, 143
Air electronics officer, 233
Air-to-air armament, 77
Air-to-surface missiles, 81–2
Airborne radar, 20, 54, 56, 61, 65, 73
Aircraft, 15–26
    Cost, 21, 26
    Crew experience, 24
    Endurance, 4, 6–7, 15–16, 18–20, 25, 27, 29, 34–7, 46, 51–2, 56, 73, 174, 192–3, 197–8, 203
    Intercommunications system (ICS), 174
    Mobility, 5, 10, 15, 22–3, 76, 174, 216
    Protection against attack, 15, 194
    Sensor coverage, 9, 15, 17–18, 225
    Wing design, 16, 18
Aircraft models:
    AP-3C Orion, 177
    B-24 Liberator, 22, 26
    D.H. 6, 58
    Nimrod MR2, 74, 236–7
    Nimrod MRA4, 236
    P-1, 71
    P-2 Neptune, 177
    P-3A Orion, 177
    P-3B Orion, 177
    P-3C Orion, 26, 71, 75, 177, 222, 230, 236

P-8A Poseidon, 178
S-3A, 236
Short Sunderland, 22
Tu-142, 74
Y-8Q, 71
Aircraft powerplants:
    Jet engines, 19
    Piston engines, 18–19
    Turbofan, 19
    Turboprop, 19
Algorithms, 72, 85, 104, 199–200
Ambient Noise (AN), 95, 147
Ambient Noise Limited, 100–01
Ambient noise spectrum, 137
Ambient noise types:
    Biological, 98
    Shipping, 98, 138
Annulus, 129, 143, 211
Anti-aircraft artillery, 21, 85
Anti-personnel weapons, 77, 81
Anti-ship cruise missiles, 2
Anti-ship missile, 82
Anti-Surface Improvement Program, 26, 177
Anti-surface warfare (ASUW), 44, 50, 82
Area of Probability (AOP), 105
Area Tracking, 223, 235
Arming and fusing assembly, 78
Arrival path type:
    Bottom bounce, 126–8, 131–2, 134–6, 149, 151–5, 216
    Convergence zone, 64–6, 74, 126, 128–30, 132–6, 143–4, 149–50, 152–4, 200, 211, 216
    Direct path, 126–8, 130, 135–6, 149–55, 211, 249
    Reliable Acoustic Path, 67, 130
Ascension Island, 131
Asdic, 50
Aspects of Bayesian search:
    Confidence levels, 195, 205, 207
    Posterior distribution, 195, 199–200
    Prior distribution, 195–6, 200
    Probabilistic weights, 196
    Probability functions, 196
Aspects of Full Spectrum anti-submarine warfare:

Choke Points, 2, 7, 67, 82–3, 169, 194, 198
Defeat shore-based C2, 2, 4
Kill boxes, 2, 10
Open ocean, 8–10
ASW kill chain, 76, 173, 187, 205, 238
ASW Operations Group (ASWORG), 54
Atmospheric ducting, 141
Atmospheric visibility, 142
Attack carriers, 63
Attenuation, 107, 114, 116, 124, 148
Auxiliary intelligence gathering ships (AGIs), 171
Auxiliary systems, 95

Background noise, 61, 95, 97–8, 107, 137, 217
Baltic Sea, 108
Barents Sea, 171
Bathymetric features, 99, 108, 130, 133, 135, 144, 146, 148, 197
Bathymetry, 6–7, 98–100, 112, 131–2, 135, 151–4, 189, 194, 197, 211, 216, 243
Battle of the Atlantic, 44, 52, 56, 74, 165, 167, 169
Bay of Biscay, 13, 55–6, 58–9, 167, 171, 203
Bayesian Inference, 195, 203
Bayesian search, 195–6, 201, 203
Benefits of convoy operations, 48–9
Bernoulli hump, 108–09
Bi-static angle, 111
Biological sound sources, 140
Bioluminescence, 93, 109–10
Blanking zone, 105
Bletchley Park, 53, 165
Blockade, 47
Body language, 176
Bogue (CVE-9), USS, 167
Bombs, 47, 51, 77–80, 82–4, 90, 240–2, 246, 249
    500-pound ASW bomb, 79
    dual-action fuses, 79
    explosive material, 78, 85
    explosive power, 79, 81, 83, 250
    frag patterns, 78–9
    frangible case, 78
    specialized fuses, 78–9
Bombsight, 78
Bottom bounce, 126–8, 131–2, 134–6, 149, 151–5, 216
Bottom losses, 128, 130–1, 134, 149
British refusal to adopt convoys, 48
British World War I ASW efforts:
    Dover Barrage, 48
    Scarecrows, 48
    Use of convoys, 49, 165

Broadband noise, 61, 89, 96–7, 244
Bubble generator, 89
Buoy washover, 141
Butterfly shape plot, 101

Cable strumming, 98
Carrier wave, 97
Cases of Localization, 205, 218
Caustic, 129
Cavitation, 34, 84, 96, 150–1, 193, 212
Central Powers, 46–7, 50
Circulation models, 171
Clarity of seawater, 140
Close Tracking, 223, 227, 229, 231–3
Cold eddies, 136
Cold War, 2, 7, 13–14, 26, 28, 41–4, 57, 60–1, 63, 65–71, 73–5, 81, 86, 90–1, 100, 111, 126, 128, 130, 168–72, 202–03, 221, 226, 250
Command and control (C2), 1
Commander's intent, 238, 246
Compartmentalization, 158
Computer assisted search (CAS), 64–5, 68–9, 195–6, 199–200, 202–04
Confidence levels, 195, 205, 207
Conjugate depth, 133–5
Continental Shelf, 29, 110, 131–2, 135, 153
Continental Slope, 131–2, 134, 144
Convergence zone, 64–6, 74, 126, 128–30, 132–6, 143–4, 149–50, 152–4, 200, 211, 216
Convergence zone features:
    Annulus, 129, 143, 211
    Caustic, 129
    Conjugate depth, 133–5
    Critical depth, 129, 133–5, 152
    Depth excess, 129–30, 132–5, 143–4, 149, 152–3, 211
    Probability of Convergence Zone formation, 129, 135, 144
    Reswept zone, 128–9, 134, 143, 150
    Sound speed excess, 129–30, 132, 143–4, 149, 152
Convoy systems, 49, 52–4, 165–6
Coolant pumps, 42, 96–7, 150–1
Counter detection, 226, 230, 233–5
Coverage factor, 191–2, 199, 202
Crew Resource Management (CRM), 173, 223
Crew stability, 185–6
Crew training, 39, 183
Crew-served weapons, 76
Critical depth, 129, 133–5, 152
Cueing information, 5, 7, 9, 11–12, 15–16, 31, 61, 64, 66–7, 109, 156, 158–9, 164,

167–8, 170–1, 189, 192, 196–7, 199, 201, 203, 222
Cueing systems, 159
Cueing types, 160
　Acoustic Intelligence (ACINT), 155
　Communications intelligence (COMINT), 161
　Cryptanalysis, 42, 58, 161–3, 165, 167
　Datum Cueing, 164
　Imagery Intelligence (IMINT), 160
　Signals Intelligence (SIGINT), 12
Cumulative detection probability (CDP), 104, 188

Decibels (dB), 95
Deck-mounted guns, 86
Deep Scattering Layer (DSL), 117
Deep Sound Channel (DSC), 9, 62
　DSC axis, 126, 136
Deep-water TL plot, 149
Defensive minefields, 83
Depth charges, 41, 77, 79–81, 84, 90, 240, 249–50
Depth excess, 129–30, 132–5, 143–4, 149, 152–3, 211
Detection Area, 153–4
Diesel-electric hunter killer submarine (SSK), 2
Dinoflagellates, 109
Dipolar spreading, 115
Direct blast, 105
Direct Path, 126–8, 130, 135–6, 149–55, 211, 249
Directional Variation in Transmission Loss, 152
Directivity Index (DI), 95, 99, 101, 147, 149
Diurnal temperature variations, 136
Diurnal variation, 117
Doppler effect, 88, 102
Downslope Enhancement, 132

East Pacific Rise, 140
Echo repeater, 89
Eddies, 135–6, 216, 224, 232
Eddy, 126, 136–7, 216, 235
Effectiveness of bottom mines, 83
Electro-optical (EO) sensors, 3
Electronic support measures (ESM), 4
Electronic warfare operator, 179
Employment rhythm, 181–2
Engineering lineup, 150
Enigma cipher, 53
Escort carriers, 55, 63
Escort operations, 193
Event Tracking, 223, 227, 229, 232–3, 235

Expendable bathythermographs (XBTs), 120
Explosive Echo Ranging (EER), 68

False alarm rates, 71, 104
Far field waves, 109
Farthest-on circle (FOC), 197–8
Feather, 92–3, 205
Fiber-optic cables, 164, 172
Fighter aircraft, 23, 26, 50, 77, 82
Figure of Merit (FOM), 99–100, 143, 148–52, 154–5
Fixed Deployable System (FDS), 67
Fixed function tactical systems, 175, 177–8
Flotation bag, 98, 226
Forward-firing weapons, 81
Frag pattern, 78–9
Fragmentation, 78–80
Fratricide, 245
Frequencies of interest, 97, 137–8, 220
Frequency-modulated (FM) pulses, 102
Fronts and eddies, 135, 232
Full Field plot, 152
Full Spectrum ASW, 1, 13–14
Fuse types:
　Acoustic, 83
　Contact, 85
　Dual-action, 79
　Faulty, 79
　Hydrostatic, 80
　Influence, 83
　Magnetic, 83, 85
　Shallow-water, 80
　Specialized, 78–9
Fuse-delay mechanisms, 79
Fuses, 51, 73–80, 83, 85, 240, 248

General-purpose noisemaker, 89
Geomagnetic activity, 217
German tanker submarines, 55
GIUK gap, 67, 74, 108
Global Positioning System (GPS), 218
Gravity waves, 124

Hawaiian archipelago, 131
Hawaiian-Emperor seamount chain, 131
High frequency direction finding (HF/DF), 39, 56, 164–8
High radiated noise levels, 20, 62, 130, 132
Holland, John, 45
Human-machine teaming, 73, 104
Humidity, 114, 140–2
Hydrophones, 28, 49, 54, 61–2, 67, 94, 98, 105, 128, 137–8, 143, 163, 169, 219

In-situ environmental data, 100
Indications and warnings (I&W), 156–7, 160, 162, 171

Information management, 173–4, 177–8
Instantaneous detection probability (IDP), 188
Integrated Undersea Surveillance System (IUSS), 170, 172
Intelligence gathering, 62, 64, 111, 156–7, 159–63, 165, 167, 169, 171, 205
Internal wave detection, 73
Internal waves, 66, 108, 124–5
International law, 46
    Commerce raiding, 46
    Enemy merchant ships, 46
    Neutral merchant ships, 46
Inverse synthetic aperture radar (ISAR), 94
Ionic relaxation, 116, 142
IR guidance systems, 86
Isovelocity condition, 118

Japan Maritime Self-Defense Force (JMSDF), 71

Knudsen, Vern, 144
Knudsen Spectra, 138, 144, 148
Kola Peninsula, 169, 171
Kriegsmarine, 166–8
Kurier, 56

Lattice crystal structure, 116
Leigh Light, 54
LIDAR, 66, 73–4, 107–08, 112
Lightweight torpedo (LWT), 42, 54, 68, 83–5, 90, 146, 155, 232, 242, 244, 246–7
Lightweight torpedo components:
    LWT seeker head, 155
    Shaped charge, 85
Limiting ray, 123–4
Lithium ion batteries, 73
Load sharing, 175
Localization patterns, 146, 210, 215, 217
Localization plan, 234
Localization tactics, 205, 210–15, 217–18
LOFARgram, 96, 98
Long-range detection, 74, 126, 130, 135, 169, 207
Long-range propagation phenomena, 123, 146
Long-range propagation phenomenon types:
    Deep sound channel, 9, 62, 125
    Half channel, 122–3, 125
    Internal sound channels, 123, 126
    Surface duct, 123–5
Low-altitude delivery, 79
Low-Frequency Analyzer-Recorder, 96

Machine guns, 76–7
Magnesium sulfate, 116, 142
Magnetic Anomaly Detection (MAD), 54, 57, 60, 66, 73, 92, 106–07, 112, 160, 202, 217, 248
    Introduced, 27, 43, 54, 67, 71, 81, 90, 190, 236
    Long-range MAD equipment, 73
    Magnetic storms, 106
    Radio transmissions, 106–07
    Solar activity, 106
    Soviet use of, 66
Man Portable Air Defense Systems (MANPADS) 86–7, 90
Mark on top, 226
Measures of effectiveness, 146, 154, 191
Mediterranean Sea, 71, 125, 143
Mentorship, 183–4
Meteorological forecasts, 138
Mid-Atlantic Ridge, 131, 140
Minefields, 35, 43, 48, 50, 82–3
Mine types:
    Bottom, 83–4
    Moored, 48, 83, 160
Mines, 5–7, 48, 82–3, 108, 144
Mission kill, 7, 238–9
Mixed layer, 121–5, 128, 132–6, 149, 151, 211
Mixed layer depth, 123–5, 133, 136, 211
Mobile countermeasure types:
    MG-74 Korund, 91
    Mk 70, 91
Mobile countermeasures, 89–90
Mobility kill, 238–9, 243
Modeling adversary behavior, 156
Modeling methods:
    Bayesian Inference, 195, 203
    Monte Carlo, 189
Monostatic active sonars, 103–05
Multispectral imaging, 73, 160
Multistatic Active Coherent (MAC), 71
Multistatic Active Sonar, 73, 75, 103
Multistatic active sonobuoy fields, 103
Multistatic sonobuoy fields, 103, 112

Narrowband sound sources, 96
NATO, 13, 25–6, 41–2, 60–8, 90, 94, 163, 168–71, 250
NATO undersea surveillance program, 169
Need to know, 158
Next Generation Airborne Passive Sensor, 72
Noise Level (NL), 98–9, 101–02, 147
Noise-damping, 28
    Anechoic tiling, 28, 102

Noise-damping isolation, 28
Propeller design, 66
Rafts, 28, 63
Noisemaker, 89
Non-acoustic sensor methods, 73
Norwegian Sea, 26, 67, 71, 171
Nuclear explosions, 81
Nuclear weapons, 44, 60, 81

Observe, Orient, Decide, and Act (OODA)
    loop, 173
Ocean analysis, 135
Oceanographic analysis, 185
Offensive mining, 82
Oil and gas extraction, 139
Operator consoles, 174, 176–7, 179
Optically guided, 82

Passive CDP, 189–90
Passive sonar, 4, 28, 30–1, 34–5, 38, 54, 56,
    61–2, 69, 73, 84, 90, 94, 98–102, 106,
    111, 140, 147, 149, 155
Passive sonar equation, 99, 101, 147, 149,
    155
Patrol boxes, 55, 63, 65
Periscope, 5, 9–10, 12, 17, 22–3, 29–30,
    32–3, 36–7, 40, 47, 61, 87, 92–4, 141,
    151, 164, 191–2, 199, 205–06, 209, 211,
    217, 235, 237
Periscope depth, 5, 22, 37, 40, 61, 92, 151,
    206, 211, 217, 235, 237
Periscope-detection radars, 9, 94
Ping plans, 199
Pinnacles, 84, 216
Plankton, 110
Plot stabilization, 225–6, 242, 246
Posterior distribution, 195, 199–200
Prairie-Masker, 11, 14
Pressure hull, 41, 78, 85, 101
Prior distribution, 195–6, 200
Probabilistic weights, 196
Probability of Convergence Zone
    formation, 129, 135, 144
Probability of detection, 58, 154, 187–9,
    197–8, 200, 202
Probability of kill, 232, 238–9
Probability of radar detection, 191
Propeller shaft seal, 239
Prosecution plans, 73

Radar cross section, 82
Radar performance, 140–1, 191
Radar warning receiver (RWR), 55
Radio carrier frequency, 97
Radio direction finding, 164

Radio transmissions, 13, 16, 53, 107, 161–2
Radio wave refraction:
    Standard refraction, 141
    Sub-refraction, 141
    Super refraction, 141
Range gate blanking, 103
Rate of sensor expenditure, 20, 225–7, 231,
    233
Ray, 51, 75, 104, 118–19, 123–4, 134,
    146–7, 151–2, 154, 200, 211
Ray path, 118, 147, 154, 200, 211
Ray trace, 119, 124, 147, 152
Ray tracing, 118–19, 124, 147, 152
Rayleigh parameter, 142
Reactor coolant pumps, 150–1
Recognition Differential (RD), 97–9,
    101–02, 147, 149
Reconfigurable tactical systems, 174, 178
Reconnaissance satellites, 159–60, 172
Reflection, 88, 119, 122–3, 127–8, 131,
    146, 170
Refraction, 88, 117–18, 122–3, 127, 130,
    141, 146–7
Reliable Acoustic Path (RAP), 67
Remotely-operated vehicles, 164
Reswept zone, 128–9, 134, 143, 150
Reverberation, 81, 100, 102, 117
Reverberation Limited, 100, 102
Revisit time, 192, 227, 229
Route navigator, 233
Royal Air Force, 59
    Bomber Command, 51, 53
    Coastal Command, 52–4, 59, 74
    Fighter Command, 53

Salinity, 27, 114, 120–2, 126, 135, 143–4
    Salinity level, 120
    Salinity variations, 122
Sandbag syndrome, 180
Scattering, 98, 112, 116–17, 119, 127–8,
    131, 142, 148
Sea control, 46, 57, 60
Sea of Okhotsk, 171
Sea state predictions, 141
Sea states, 141, 199
Sea water across the hull, 96
Seafloor sensors, 159, 164
Seaquakes, 139–40
Search areas, 64, 73–4, 154, 158, 168, 171,
    187, 192–3, 198–9, 211
Search cueing, 158, 170
Search planning, 14, 146, 154, 156, 159,
    187, 195–6, 200, 202, 215
Search resources, 192
Seaworthiness kill, 7, 238–9, 243, 250

Security of sources and methods, 157–8
Self Noise (SN), 98
Sensor coverage, 9, 15, 17–18, 225
    Altitude, 9, 15, 17–19, 76, 78–9, 141,
        159, 174, 213, 236, 246–7
    Field of view, 9, 17, 33, 38, 47, 88, 93,
        103, 110, 164, 191, 210
    Line-of-sight, 17, 40
    Search rates, 4, 9, 17–18, 33, 42
Shadow zone, 124
Shallow-water ambient noise, 138
Shaped charge, 85
Shelf Break, 131–2
Shipwreck, 104, 206–07
Signal characteristics, 205
Signal Excess (SE), 99–102, 144, 147–50
Signal processing, 33, 61–2, 68, 71, 74, 88,
    92, 96–7, 104, 109, 236
Signal to Noise Ratio (SNR), 95, 134, 144
Signals Intelligence, 12, 161
    Bletchley Park, 53, 165
    Enigma cipher, 53
Snell's Law, 117–19, 141, 147
Snorkeling, 36–7, 61, 73, 94, 191
Sodium chloride, 142
Solar radiation, 123, 136, 143
Sonobuoy barrier, 198, 211
Sonobuoy drop points, 146
Sonobuoy types:
    Active, 100, 102–03, 112, 190, 194, 201,
        207
    Deep-drifting, 72
    Long-dwell passive, 72, 236
    Multistatic, 72, 190
    Passive, 10, 72, 97–8, 105, 178, 202, 207,
        209, 215, 217, 222, 233, 236
Sound Pressure Level (SPL), 95, 130, 137,
    149–50
Sound propagation, 27, 64, 105, 114–16,
    118–19, 122, 124–5, 130–2, 135–6, 140,
    146–7, 149, 151–3, 155, 169, 185, 216,
    224, 232
Sound propagation paths, 115–16, 119,
    122, 152, 169
Sound refraction, 117, 146–7
Sound speed excess, 129–30, 132, 143–4,
    149, 152
Sound speed minimum, 122, 125–6
Sound Speed Profile (SSP), 99–100, 111,
    119–23, 125–6, 128–30, 132, 143–4,
    147–54, 207, 211, 216–17, 224, 232, 242,
    246
Sound surveillance system (SOSUS), 42,
    62–3, 65–7, 74, 169–72
Sound wave reflection, 119

Source Level (SL), 95, 99, 101–02, 147–50,
    153, 188, 216
Source sonobuoys, 103–04
Soviet Union, 26, 42–3, 57, 60, 63, 65,
    68–70, 84, 112, 170, 237
Spall damage, 85
Special weapons, 81
Specular return, 104, 111
Spreading loss, 115–16
Spreading mechanisms, 115
    Cylindrical spreading, 115, 126
    Dipolar spreading, 115
    Spherical spreading, 115
Standardized tactical employment, 184,
    186
Static countermeasures, 89
Steps of ASW kill chain:
    Attacking, 3, 8, 13, 22–3, 25–6, 30, 34,
        40, 46, 49, 51, 53, 60, 76–7, 80–1,
        84–6, 164, 166, 193–4, 211, 213–15,
        219, 238, 244–5, 247
    Localization, 207–22
    Mission planning, 205
    Search, 187–204
    Tracking, 223–7
Steps of the Attack, 244
sticks, 80
Strait of Gibraltar, 143
Submarine classes, 42
    Ohio class, 108
    Project 661/PAPA class, 74
    Project 671RTM/VICTOR III, 74, 112
    Project 675/ECHO II class, 231–2
    Project 685/MIKE class, 74
    Project 945/SIERRA, 91
    Project 949/TYPHOON, 90–1
    Project 971/AKULA, 74, 112
    "Scorpène" class, 70
    Seawolf class, 42
    Type 209, 70
    Type 214, 70
    Type XXI, 56
    Västergötland class, 70
Submarine communications:
    Data transmission, 39
    Extremely low frequency, 39
    Floating communications buoys, 40
    Message traffic, 40, 53, 66, 165–6
    Satellite communications, 40
Submarine countermeasure systems, 76
Submarine evasion, 88, 135, 244
Submarine hull types:
    Circular, 45
    Double, 41, 45, 67–8, 84
    Hydrodynamic outer, 85